Alfred Price

Instruments of Darkness

The History of Electronic Warfare

The instruments of darkness tell us truths,
Win us with honest trifles, to betray's
In deepest consequence.

MACBETH

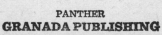
PANTHER
GRANADA PUBLISHING
London Toronto Sydney New York

Published by Granada Publishing Limited
in Panther Books 1979

ISBN 0 586 04834 0

Revised edition published by
Macdonald and Jane's Publishers Limited 1977
Copyright © Alfred Price 1967, 1977

Granada Publishing Limited
Frogmore, St Albans, Herts AL2 2NF
and
3 Upper James Street, London W1R 4BP
1221 Avenue of the Americas, New York, NY 10020, USA
117 York Street, Sydney, NSW 2000, Australia
100 Skyway Avenue, Toronto, Ontario, Canada M9W 3A6
110 Northpark Centre, 2193 Johannesburg, South Africa
CML Centre, Queen & Wyndham, Auckland 1, New Zealand

Set, printed and bound in Great Britain by
Cox & Wyman Ltd, London, Reading and Fakenham
Set in Intertype Times

Granada Publishing ®

FOR JANE

Foreword

by

Sir Robert Cockburn, KBE, CB

World War II was dominated by air power, which permeated every phase of the conflict. The aircraft became a major instrument of offence and defence in the air and a vital weapon in support of ground forces, in maritime warfare, in reconnaissance and in transport. This expansion of air power was stimulated by, and became critically dependent on, a series of remarkable developments in the fields of radar and radio communications. Both sides committed large resources to successive systems of early warning and detection, navigation, target identification and weapon guidance. Under the stimulus of war technology advanced rapidly and each new system provided greater range, greater precision and greater capacity. Yet by modern standards they were still relatively naïve in concept and were soon found to be vulnerable to interference, deception and manipulation. It was a rude shock to designers to discover how quickly performance demonstrated in the laboratory was nullified in operation against a resourceful enemy; and as the war progressed scientists and engineers pitted their wits against one another to preserve their own systems, and to discover and exploit the weaknesses in those of their opponents'. It is this story which Alfred Price describes in *Instruments of Darkness*.

Alfred Price was a serving officer in the Royal Air Force for a time with Bomber Command, and he has been able to recapture the excitement and drama of a struggle in which new techniques and tactics could have such immediate and catastrophic consequences. But he is also an electronics specialist well qualified to deal with the technical aspects of his subject and to appraise the relative importance of the various countermeasures during and since World War II.

Rarely before or since has it been possible for the scientist in the laboratory to make such a direct impact on military operations. A few black boxes based on a new piece of intelligence, a revealing reconnaissance photograph or observations by a returning bomber crew could within a few weeks or months affect the fate of cities and the lives of hundreds of aircrew. Not all black boxes were equally effective; some were at best of psychological value, some were a temporary nuisance, and some were not only useless but positively dangerous. In the heat and fog of war any opportunity, however slight, must be exploited, but in retrospect it is clear that clever devices and adroit tactics were usually of limited value. Trivial weaknesses in a system were easy to exploit but equally easy to remedy, and over-sophisticated jamming and warning methods were incompatible with the nightly holocaust over German targets. Subtle countermeasures like 'Moonshine' were effective for special operations where surprise could be exploited, and they played an important part in the spoof invasions of the Pas de Calais. But it was straightforward noise jamming and the massive use of 'Window' which was most effective in sustained operations.

Alfred Price has profited by the lapse of thirty years to put his story of World War II into perspective. He has gone to a lot of trouble to present at each stage both the British and the German story, and he shows how closely developments on one side were matched on the other. There was for example the extraordinary similarity in the evolution of British 'Window' and German *Düppel*. Nowadays it is accepted that major technical advances will occur almost simultaneously in a number of countries, even in highly classified military projects. In peacetime the time scale of development is long enough that a year either way in producing a new weapon may not seriously affect the issue, but in war six months can make the difference between victory and defeat; and it was by such narrow margins that the outcome of the Radio War was determined.

Both sides entered the war believing that they possessed in radar a unique advantage over the other; and neither foresaw the profound effect that this scientific breakthrough would have on air operations. In 1940 our own radars were in many respects inferior to the *Freya* and the *Würzburg*, and we had no bombing and navigation systems to compare with the *Knickebein* beams and their successors. At the end of the war the Germans were introducing, well ahead of the Allies, a range of guided weapons, including the pilotless aircraft V-1 and the ballistic missile V-2. But the German High Command did not properly appreciate the pace of development or its inevitable impact on operations. For two critical years they failed to maintain the momentum of research in radar, and rapidly lost their initial advantage. By the end of the war British and American equipments were far superior both in performance and in range of application; and the German guided weapons came too late to redress the balance. In particular we were quicker to recognize and exploit the intrinsic vulnerability of radar and radio systems, and the initiative in the jamming war lay firmly in our hands.

The importance of the electronic environment not only to aircraft and guided weapons but to the whole range of military operations is now well understood; and one of the most important criteria of any radar or radio system is the ability to discriminate against unwanted and irrelevant information. Vulnerability can be theoretically specified, and allowed for. Nevertheless, jamming and deception are always possible, with sufficient effort. Ideally an economic balance should be struck between complexity and vulnerability, so that the cost of nullifying a system is comparable to the cost of establishing it. But this condition can seldom be met in practice, and Alfred Price's book is a salutary reminder that the impressive panoply of modern weapons is dependent ultimately on the survival, in war, of their guidance and control systems.

Introduction to the Second Edition

The first edition of *Instruments of Darkness* appeared in 1967 and described the story of electronic warfare during the Second World War. Revised ten years later, this second edition contains two additional chapters to bring the story up to 1977. Unfortunately security considerations have prevented the final two chapters from being as detailed as those earlier. Perhaps one day it will be possible to put together both sides of the story of the air wars over North Vietnam and the Middle East, but that day is a long way away.

In *Instruments of Darkness* I have done my best to produce an account of the electronic war which can be understood by those not directly concerned with this field. It is far too important a subject for descriptions of it to be confined to popular journalism on the one hand, or jargon-ridden technical prose on the other. If it is possible to steer a middle course between the two, I hope I have done so.

Uppingham   ALFRED PRICE
Rutland, 1977

Contents

List of Illustrations

Acknowledgments

In writing *Instruments of Darkness* I have been greatly helped and encouraged by many busy men who have unhesitatingly given me their valuable time: to all of them I tender my grateful thanks. Space does not allow me to mention each one by name, but I am particularly indebted to Sir Robert Cockburn, Professor R. V. Jones, Professor D. A. Jackson, Air Marshal Sir Robert Saundby, Air Vice-Marshal E. B. Addison, Air Commodore Chisholm, Dr B. G. Dickins and Mr. J. B. Supper and, in Germany, the *Studiengruppe der Luftwaffe*, the Telefunken Company and Herr Hans Ring.

I should like to thank Sir Donald MacDougall for allowing me access to Lord Cherwell's papers and Mr R. Bruce and Mr C. Moore for allowing me to examine documents under their control. Also I am indebted to Mr L. A. Jackets and the staff of the Air Historical Branch for much of the material I have used. I must stress, however, that I alone am responsible for the opinions expressed.

My thanks go to Her Majesty's Stationery Office for permission to quote from *The Strategic Air Offensive Against Germany 1939–1945* by Sir Charles Webster and Noble Frankland; this is the 'official history' referred to in the text. I should also like to thank Messrs Cassell and Co. for allowing me to quote from Winston Churchill's *The Second World War;* Messrs Methuen and Co. for permission to quote from *The First and the Last* by Adolf Galland; Group Captain J. R. D. Braham for the use of the passage from his autobiography *Scramble*; the proprietors of the *Daily Mirror* for permission to reproduce the Buck Ryan cartoon strip; Lieutenant-Colonel Richard Milburn of the US Embassy in London for invaluable help with material and photographs; Mr Pook of Chemring Ltd; Jim Gregory of the Boeing Company; and *Air Force Magazine*, published by the US Air Force Association for permission to quote parts of Captain Carson's article 'Flying the Thud'.

I should like to record my debt to my wife, whose encouragement was so welcome when the going was tough, and my mother, who helped with the German translations. Lastly, but most of all, I should like to record my indebtedness to my good friend and severest critic, David Irving, whose unstinting help and sound advice were invaluable.

Prologue

One evening in December 1939, the German pocket battle-ship *Graf Spee* was scuttled in international waters in the estuary of the River Plate, off Montevideo. Five days before, she had suffered several hits in a battle with three British cruisers and her captain thought it unlikely that she would be able to fight her way back to a friendly port. But the River Plate has a shallow estuary, and when the demolition charges blew the ship's bottom out, she sank only ten feet before coming to rest on the sea bed. At first light next morning, a fleet of small sightseeing boats put out from Montevideo to look at the shattered warship, which was still on fire. Scores of photographs were taken of the wreck, and the pictures were flashed round the world by news agencies.

Most people failed to notice one strange feature on the dramatic close-up photographs of the blazing hulk: an aerial-array looking rather like a bedstead on its side was mounted above the battleship's bridge. British Intelligence sent its own sightseer to look at the battleship on their behalf: Mr L. Bainbridge Bell, a radar expert, arrived in Montevideo soon after. He boarded the wreck and climbed up to the aerial-array – a feat requiring some agility since the battleship had developed an uncomfortable list. Bell reported to London that the strange aerials certainly belonged to a radar set, probably for ranging in the ship's guns.

Armed with this information, naval Intelligence officers in London examined the file of photographs of *Graf Spee*, and they observed that the same structure was present, though hidden under close-fitting canvas covers, on photo-

graphs taken as early as 1938. This was a discomfiting discovery, since the Royal Navy had no gun-ranging radar,
and would not receive its first set until 1941. However, Bainbridge Bell's report was pigeon-holed, and the scientific
Intelligence section of the Air Ministry did not hear of it
until another eighteen months had passed.

1

'Doctoring' the Beams

During the human struggle between the British and German air forces, between pilot and pilot, between AA batteries and aircraft, between ruthless bombing and the fortitude of the British people, another conflict was going on step by step, month by month. This was a secret war, whose battles were lost or won unknown to the public; and only with difficulty is it comprehended, even now, by those outside the small high scientific circles concerned.

— WINSTON CHURCHILL, *Their Finest Hour*

The truth about Intelligence work in war is that much of its success depends on chance, and much upon tenacity; and little of it is glamorous in the way that readers of modern espionage thrillers would believe. Intelligence officers have several ways of getting information on enemy developments, and the spy is just one of these. Indeed, it is often the mundane methods that produce the best results. In the case of a secret device to guide bombers to targets, for example, it was usually only a matter of time before a raider fitted with it was forced down and captured. Then examination of the wreckage would reveal its secrets. Prior to an attack crews could not reasonably be expected to memorize long lists of radio frequencies, call-signs and the geographical positions of beacons; if this vital information was to be used in the stress of action it had to be written down, and taken on the sortie. And sooner or later these briefing sheets were bound to be captured. Survivors of the crash, perhaps still shaken after a narrow escape, might be induced to talk by their interrogators.

Moreover, if radio beams were used the Intelligence services had one clear advantage from the outset: such beams cannot be concealed. One has only to look and they can be found; and once they are found, their nature can be analysed and their purpose deduced. Thus a handful of men, working hundreds of miles from the enemy, can cause the whole strategy of war to change.

This was how, a little after 10 p.m. on 21st June 1940, a lone Anson aircraft came to be patrolling the skies over East Anglia: in its rear the wireless operator was carefully searching the ether with a radio receiver. Suddenly he found what he was looking for – a series of dots, sixty to the minute, piercingly clear in his headphones. As the Anson flew on, the dots were heard to merge into one steady note; a little later, the steady note broke up – not into 'dots' but into 'dashes' at the same steady rate of sixty to the minute.

The aircraft had flown through a radio beam emanating from Germany. The beam was no death-ray. Indeed, if its energy had been applied to a thimbleful of water for two hours, it would not have raised its temperature by so much as one-millionth of a degree. None the less, it posed a formidable threat: it was set to guide German bomber formations to the Rolls-Royce factory in Derby, and was techni-

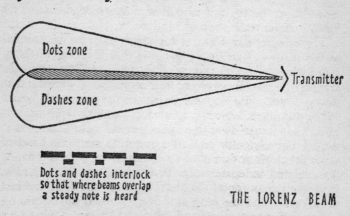

Key:　▨▨▨▨ *Steady note zone*

Dots zone

Dashes zone

〉Transmitter

Dots and dashes interlock
so that where beams overlap
a steady note is heard

THE LORENZ BEAM

cally capable of doing so with great accuracy even on the darkest of nights.

Early in the 1930s, the Lorenz company had developed a blind-approach system to help civil aircraft find landing grounds in bad weather; the system used two adjacent radio beams to mark out a path extending a distance of up to thirty miles from the airfield. In the left-hand beam Morse dots were received, and in the right-hand one dashes. The signals interlocked, so that where the beams overlapped a steady note was heard: aircraft then navigated by flying down the steady-note zone until they came to the beams' transmitter.

By the mid-1930s, the 'Lorenz' system was widely used, not only in civil airlines but in some air forces. The Royal Air Force ordered it, as did the German Air Force. In 1933 a German specialist in the propagation of radio waves, Dr Hans Plendl, started work on the adaptation of the Lorenz system to help aircraft bomb accurately. Within five years, this work had begun to bear fruit: the system developed by Plendl was called '*X-Gerät*' – the X-device. It employed a number of Lorenz-type beams: one, the approach beam, pointed straight at the target, far away; the others crossed it at three points in front of the bomb-release point. All the radio beams were transmitted on frequencies between 66 and 75 megacycles per second.

The bomber flew towards the target on a course parallel to the 'approach beam'. When still fifty kilometres (thirty miles) from the bomb-release point, the aircraft flew through the first of the cross-beams; this served as a warning that it was time to line up accurately in the approach beam. When twenty kilometres (twelve miles) from the release point the aircraft flew through a second cross-beam. As it did so, the navigator pressed a button to start one hand of a special clock not unlike a stop-watch, but with two hands that rotated independently. Five kilometres (three miles) from the release point the aircraft passed the third and last cross beam, and the navigator again pressed the button on his special clock: the hand which had been moving now stopped, and the other hand started rotating to catch it up. In fact, the distance from the second cross-beam to the third cross-beam was three times that from the third cross-beam to the bomb-release point (three miles), so this second hand travelled three times faster than the first. When both hands coincided, a pair of electrical contacts was closed, and the bombs were *automatically* released.

All in all, this was a frighteningly sophisticated device, considering that the Second World War had not even begun; the combination of clock and beams provided accurate data on the bomber's speed over the ground, one of the most important facts to be known for precision bombing once the aircraft was routed accurately over the target. But the aircraft had to fly straight and level on the approach beam for the last twelve miles in to the release point, and these three minutes could be a long time for an aircraft to be committed to a dead straight flight path over enemy territory.

Because of its complexity, and the need for especially accurate flying during the bombing run, the *X-Gerät* called for specially trained aircrew. In consequence, the German Air Force established an experimental flying unit during the winter of 1938, a few weeks after the *démarche* at Munich. This unit, No. 100 Air Signals Battalion,[1] was to ready the device for service use and to train crews to use it; it was equipped with Junkers 52 and Heinkel 111 bombers, and was based at the Air Force's signals establishment at Köthen, near Dessau. Now that the *X-Gerät* was in service, Plendl's work was over: he shifted his attention to the development of radio devices to enable aircraft to drop their bombs with even greater accuracy, and at longer ranges. His *X-Gerät* had a maximum range of only about 180 miles, and allowed a mean bombing error of a hundred yards. He felt he could improve on both these figures.

While Plendl had been putting the finishing touches to the *X-Gerät* Telefunken – one of Germany's bigger electronics

1. *Luftnachrichten-Abteilung 100.*

firms, and a competitor of the Lorenz company – was working on a second blind bombing system for the German Air
Force. Their method was called '*Knickebein*' – the Crooked
Leg. Compared with its rival it was the essence of simplicity:
it employed just two Lorenz beams, one for the approach
and one which crossed at the target. Inevitably, the system
lacked the great accuracy of Plendl's device, but it had
two advantages: firstly, that any crew could use it without
special training, and secondly that it worked on the same
three frequencies as did the standard Lorenz blind approach
receivers carried in all German heavy bombers – 30, 31·5
and 33·3 megacycles per second – so the system's signals
could be received without the installation of any specialized
equipment.

The range at which this new system's signals could be
picked up increased with the altitude of the aircraft: an
aircraft flying at twenty thousand feet could receive them
270 miles away. The beams were only one-third of a degree
wide, resulting in an accuracy of one mile in 180.

By the end of 1939, the German Air Force had erected
three *Knickebein* transmitters – one at Kleve, close to the
Dutch frontier and the nearest point in Germany to England,
a second at Stollberg on the west coast of Schleswig-Holstein near the Danish frontier, and a third at Lörrach in the
extreme south-western corner of Germany, within five miles
of the French and Swiss frontiers. During the lightning
campaign against Poland, Telefunken's system could not
in fact be used as there was no time to erect the transmitters in eastern Germany. But the much smaller *X-Gerät*
transmitters were quite mobile, and they were actually
aligned on a munitions factory near Warsaw. No. 100 Air
Signals Battalion attempted a night precision bombing attack
using the beams, but the results of this operational trial
could not be observed because other bombers had attacked
the same target earlier in the day. As it was, there was no
further call for the Germans to trouble themselves with night
precision bombing; their air force held complete air supre-
macy during the day. Towards the end of November 1939,
the battalion was re-designated '*Kampfgruppe 100*' – No.

100 Squadron.[2] The equipment of *KGr100* consisted of some twenty-five Heinkel 111's, all fitted with the special X-beam receivers designed by Plendl. The unit went into action in Norway during the April 1940 campaign, but as an ordinary day-bomber squadron. It also served in this role during the campaign in France. Soon after the British expeditionary force's evacuation of Dunkirk, German Air Force troops began the erection of both *Knickebein* and *X-Gerät* transmitters along the coasts of Holland and Northern France, and these were held in readiness in case the British should be foolish enough to fight on alone: for then heavy air attacks would be needed to make them see reason.

Over a year before, the Committee for the Scientific Study of Air Defence under Sir Henry Tizard had drawn attention to the prevailing ignorance in Britain on the latest German weapons. The Air Staff had suggested that a scientist should be attached to the Air Ministry Directorate of Intelligence, and they asked Tizard to propose someone. Tizard nominated Dr R. V. Jones, a tall heavily built physicist with a brilliant mind and a keen sense of humour; the latter took up the appointment very soon after the outbreak of war, and by the first weeks of November 1939 he had already received the first remarkable clues about the German beams. On 4th November the British Naval Attaché in Oslo had received a most unusual parcel: it contained a wad of papers written in German, which appeared to discuss a number of new secret weapons. The sender referred to himself as a 'well-wishing German scientist'. The strange bequest was sent to England, where it eventually arrived on Jones' desk.

The 'Oslo Report', as it became known, was a real-life Intelligence gift, and almost stranger than anything in fiction. Its scope was vast: it spoke of the development of

2. The *Gruppe* (squadron) was the basic flying unit of the German Air Force. It usually comprised three *Staffeln* each of nine aircraft, and a *Stab* ('staff') flight of three. Three *Gruppen* made up a *Geschwader*. The bomber units were prefixed *Kampf-*, and fighter units *Jagd-*; night-fighters were prefixed *Nachtjagd-*, and heavy fighters *Zerstörer*.

large rockets, a rocket-powered gliding bomb and an important research station at a little place called Peenemünde on the island of Usedom in the Baltic. It gave the method of operation of the new magnetic torpedo and the fact – then unknown in Britain – that the Junkers 88's entering service with the German Air Force could be used both as long-range and as dive-bombers. But that was not all: there were details of a new device to enable aircrew to measure their distance from special ground stations by means of radio beams.

In general, where the information in the Oslo report could be checked it was found to be surprisingly accurate. There were two possible explanations – either the whole thing was a 'plant' designed to convince the British that Germany was technically extremely well-advanced, or else the source was genuinely disaffected with Nazi Germany and wished to tell all he knew. Jones found that where the information overlapped that already known, it was highly accurate; this pointed to the second conclusion. Others disagreed: quite apart from their general suspicion of any report painting such a bright picture of the state of German technology, it was argued that in particular no one scientist could possibly have had access to information on so many different projects. If the report appeared to be confirmed by some of the information already known in England, this was exactly what was to be expected of any good 'plant'. Only time, it was agreed, would tell.

Four months elapsed before a further clue in what was to become the hunt for German beams arrived. During March 1940, a Heinkel 111 of *KG26* crashed in England, and in the wreckage a scrap of paper was found which stated:

Navigational aid: radio beacons working on Beacon Plan A. Additionally from 0600 hrs Beacon 'Dünhen'. Light beacon after dark.
    Radio beacon *'Knickebein'* from 0600 hrs on 315°.

At about the same time, a prisoner admitted under interrogation that this *'Knickebein'* was 'something like the

X-Gerät, about which he obviously assumed the British knew. He said a beam was sent out which was so narrow that it could reach from Germany to London without a divergence of more than one kilometre. If this was so, Jones noted, the device could be used for accurate bombing at night.

Two months later, the diary of a German airman was found in the wreckage of another of *KG26*'s Heinkel 111's. Under 5th March it carried the significant entry:

Two-thirds of the *Staffel* on leave. In the afternoon we studied *Knickebein*, collapsible boats, etc.

From the snippets of information now available to him, it appeared to R. V. Jones that *Knickebein* – and the *X-Gerät* which was 'something like' it – must be directional radio beams of some sort. The given bearing of 315 degrees would point from the north-western coast of Germany to Scapa Flow, an area where *KG26* was known to have been active. What seemed unbelievable was the prisoner's insistence that a radio beam running from Germany to London – a minimum distance of 260 miles – could do so with a divergence of only one kilometre; this was far in advance of anything then possible in England. The only comfort was in the knowledge that prisoners often tell lies. In any case, assuming that the Oslo Report was genuine, it seemed strange that there should have been no explicit mention of either *Knickebein* or *X-Gerät*.

The next logical move was to examine the radio equipment carried by the Heinkel 111 – since this was the aircraft linked with the Intelligence reports on *Knickebein* – to see whether there was any device suitable for the reception of beams at long ranges: the first German aircraft to come down on British soil – a Heinkel which touched down near Edinburgh in October 1939 – had been carefully dissected at the time. Experts had analysed every single piece, and noted that its Lorenz blind approach receiver was much more sensitive than its British equivalent. Could it be that this device picked up the long-range beams?

At first sight it might seem a simple matter to find out for certain whether the Germans did have a workable system of long-range radio beacons: a few flights by an aircraft fitted with the proper equipment would soon settle the matter. But R. V. Jones was a young Intelligence officer, only recently appointed, and he had no such aircraft under his control. He had to play his cards very carefully. There could be no mistaking the gravity of the situation if the German Air Force did have an accurate method of finding their targets by night – a time when the British defences were ineffective.

In the first days of June 1940, the last of the British soldiers had been rescued off the beaches at Dunkirk. It would not be long before the Germans had established their bombers in the Calais area less than a hundred miles from London. The German Air Force had proven its destructive ability during its attacks on Rotterdam and Warsaw. Would London be next?

The evidence suggesting that the Germans had indeed perfected radio beams to help them bomb by night was certainly most nebulous: a couple of pieces of paper, a prisoner's indiscretion, and the fact that a piece of standard radio equipment used by the German Air Force differed in sensitivity from its RAF counterpart. The whole thing could still be a 'plant' by German Intelligence to divert attention from something entirely different. Dr Jones saw his position as being analogous to that of a watch-dog: he had to bark when he saw danger; but if he barked at the very first whiff of trouble and none was subsequently revealed, people would soon learn to disregard his cries. If on the other hand he barked too late, the Germans would maintain the element of surprise until they had achieved their aims. Somewhere between Scylla and Charybdis he had to steer his course.

One man could secure for Jones the influential backing he needed – a man upon whom he could rely for a sympathetic hearing: his tutor at Oxford before the war, Professor F. A. Lindemann. Lindemann was not only an extremely capable scientist, he was gifted with the ability to pass on his knowledge to those with little understanding of technical matters.

Mr Winston Churchill, for all his superlative qualities as a war leader, had little grasp of such things. He and Lindemann had been close friends since 1919, and when he now became Prime Minister in May 1940 the association continued. Professor Lindemann remained Mr Churchill's closest adviser on scientific matters and as a result held the key position in the direction of the nation's scientific war effort. Clearly, if Jones could convince Lindemann of the possible danger of the radio beams, his battle would be half won. If the hunt for the mysterious beams had Lindemann's backing, no effort would be spared to bring it to a successful conclusion.

On 12th June 1940, the Professor sent for R. V. Jones to discuss another matter. Afterwards, Jones carefully brought the conversation round to *Knickebein*. Lindemann was unimpressed, on a technical ground: he did not believe that a long-range beam on 30 megacycles would bend itself round the curvature of the earth. The employment of Lorenz beams over short ranges – up to fifty miles – was well established; but for the German Air Force to bomb targets in England using transmitters situated in Germany, the beams would have to have an effective range of over 200 miles. This was the point at issue. All the information available in England showed that radio waves on a frequency around 30 megacycles – i.e. those which could be picked up by the Lorenz receivers – did not curve round the earth's surface, but travelled in straight lines.[3] This limited their range to about 180 miles, assuming the receiving aircraft was flying at 20,000 feet; and this fell far short of the 260 mile range necessary to reach London from the nearest point in Germany.

Within ten days of this first unsuccessful encounter with Lindemann, Jones was to get the answers to most of the outstanding questions on *Knickebein*, as the result of a remarkable combination of coincidences and good fortune. In the

3. In fact partial bending to conform with the earth's curvature does occur with beamed transmissions on 30 megacycles, but this was not known in Britain at the time.

meantime, he called on Professor Lindemann again on 13th
June, and this time he took with him an unpublished paper
written by Mr Thomas Eckersley, scientific adviser to the
Marconi company and – like Hans Plendl in Germany – a
leading authority on the propagation of radio waves. The
paper contained an important series of graphs drawn
by Eckersley to illustrate the maximum ranges at which
radio signals on various frequencies could be received.
By taking the extreme end of one of the curves, there
seemed to be evidence that signals on 30 megacycles could
be picked up by aircraft flying at 20,000 feet over much
of England, if transmitted from high ground in Germany.

This satisfied Lindemann. He wrote to the Prime Minister
on the same day:

There seems to be some reason to suppose that the
Germans have some type of radio device with which they
hope to find their targets. Whether this is some form of
RDF [radar] . . . or some other invention, it is vital to
investigate and especially to seek to discover what the
wavelength is. If we knew this, we could devise means to
mislead them; if they use it to shadow our ships there are
various possible answers . . . If they use a sharp beam this
can be made ineffective.

With your approval I will take this up with the Air
Ministry and try to stimulate action.

At the bottom of Lindemann's letter, Mr Churchill jotted
a quick note before passing it to Sir Archibald Sinclair, his
Secretary of State for Air: 'This seems most intriguing and
I hope you will have it thoroughly examined.'

Now Jones had the big guns firing for him. Sinclair acted
promptly, for on the following day, 14th June, as the German army was marching triumphantly into Paris, he placed
Air Marshal Joubert in charge of the investigation. RAF
interrogators were that very day already questioning a further German airman: he stated that *Knickebein* was a
bombing device involving two intersecting radio beams

which could be picked up on the aircraft's Lorenz receiver.[4]
He added that the bombers had to fly very high to pick up
the beams at long ranges. For example, to receive the signals
over Scapa Flow the aircraft had to be above 20,000 feet.
Jones observed that from Scapa Flow to the nearest point in
German-controlled territory – western Norway – was 260
miles, the same distance from London to the nearest point
in Germany.

This new Intelligence was available in time for a meeting
called by Air Marshal Joubert on 15th June, a Saturday.
Among those present were both Lindemann and Jones. It
was agreed that the evidence was sufficient to justify bringing
more people into the picture. Joubert summoned a further
meeting the following afternoon.

At the Sunday conference, he again took the chair. Besides Dr Jones, Air Chief-Marshal Sir Hugh Dowding – the
C-in-C of Fighter Command – and Air Commodore Nutting
– the Director of Signals – were also present. Jones recounted the available evidence, and it was decided to fit special
radio receivers to aircraft allocated to hunt the beams down.
The aircraft were available within twenty-four hours – a
flight of three Ansons commanded by Wing-Commander
R. S. Blucke. Work began immediately on the fitting of the
necessary radio-receivers. One of those concerned with the
hunt for *Knickebein*, Squadron-Leader Scott-Farnie, predicted that the beams would most probably be found on a
frequency of 30, 31·5 or 33·3 megacycles, since in every
case where Lorenz receivers had been found in wrecked
German aircraft they had been tuned to one of these three
frequencies.

On that Tuesday, 18th June, fresh evidence arrived – a
miscellany of papers salvaged from a German aircraft shot
down in France some weeks earlier. On one of them was
written:

4. He also stated that the receiver automatically released the
bombs, but he was obviously confusing it with the *X-Gerät* in this
respect.

Long-range radio beacon = VHF

1. *Knickebein* (near Bredstedt, north-east of Husum)
        54° 39'
         8° 57'

2. *Knickebein*
        51° 47' 5"
         6° 6'   (near Kleve)

All this served to confirm the earlier information on *Knickebein*. The two sites seemed to have been chosen so as to be as close as possible to England, while giving the greatest possible 'angle of cut' between any pair of beams. Jones noted that Scapa Flow lay on an exact bearing of 315 degrees from Bredstedt; this explained the earlier references.

As if more proof were needed, a wrecked Heinkel 111 of *KG4* provided a further clue. The wireless operator's log had been recovered intact and it included a list of known radio beacons. But at the head of the list was a jotted entry reading, '*Knickebein* Kleve 31·5'. All the other beacons mentioned in the log were followed by a frequency, and in each case the RAF listening service was able to confirm that these were correct for the night in question. It was therefore reasonable to assume that the *Knickebein* at Kleve had that night been radiating on a frequency of 31·5 megacycles. So it seemed that Squadron-Leader Scott-Farnie's deduction would be proven correct.

One of Blucke's Anson aircraft took off to try to find the beam transmissions that same evening, but the receiver developed a fault and nothing was heard. On the night of the 20th, an Anson was again sent up to try to hunt down the beams, but again nothing was picked up: in fact the German Air Force had stayed at home. But even as the aircraft was airborne, an RAF Intelligence officer was literally piecing together the final clue: very early on the morning of the 20th, a German wireless operator had baled out of his crippled aircraft; soon after landing in England, this airman, more conscientious than many of his compatriots, realized that he still had his notebook with him, and he had carefully torn it up into three thousand pieces. But, as he had attempted to bury them, he was caught and all the bits were recovered. By 3 a.m. next morning, the bits had been pieced together again.

The reward was an untidy table of data:

V.H.F.	54 38 7″	North	} Stollberg
Knicke.	8 56 8″	East	
	51	N.	
	1 30′	E.qms. (30 mc/s)	
Cleve	51° 47′ 4	N.	
	6 2′	E.	
	55°	N.	
	2°	E. qms. (31.5 mc/s)	

This was a most important find, and served to confirm all that had gone before. 'Stollberg' was very near Bredstedt, and it seemed that the transmitter there had been operating on 30 megacycles. 'Cleve' was the old-fashioned spelling of Kleve. The other two positions mentioned on the paper were in the North Sea, but as both were in round figures they were probably turning points and could be ignored in the search.

By the morning of 21st June, R. V. Jones had therefore been able to establish with considerable precision the positions and the frequencies of two of the *Knickebein* transmitters. It was fortunate that he had done so, for that very morning Mr Churchill had summoned a top-level meeting at No. 10 Downing Street to acquaint himself with the latest Intelligence on the German radio beams.[5] Jones himself learned of the conference only when he arrived at the Air Ministry that morning. He found a note on his desk telling him to report to the Cabinet Room at No. 10, but suspected that the summons was a practical joke, perhaps in retaliation for one he had himself played. He telephoned Scott-Farnie, from whom it was said to have come, to check. The note was genuine. He hurried round to Downing Street, and

5. Among those present were Sir Archibald Sinclair, Lord Beaverbrook, Professor Lindemann, Sir Cyril Newall (Chief of Air Staff), Sir Hugh Dowding and Sir Henry Tizard.

arrived to find that the meeting had been in progress for half an hour. An argument was raging about whether aircraft could in fact be guided by long-range radio beams. Jones had not been there long when Mr Churchill questioned him on a technical point: it was the very excuse Jones had been waiting for. He recounted to the Prime Minister the mass of evidence supporting the theory that the Germans had developed a system of radio beams to direct their bombers on to targets. This seemed to convince even the most sceptical that here was something worthy of further investigation. The axis of the discussions shifted from 'Do the beams exist?' to 'How can we find out more about them?'

During the afternoon, Air Commodore Nutting summoned Jones and Eckersley to his office to discuss the technical details of the beam transmissions, if they were used by German bombers over England. It was now that Eckersley dropped his bombshell: despite the important series of graphs he had drawn, he said he could not agree with the widely-held explanation of *Knickebein*: radio signals on 30 megacycles just would not bend round the curvature of the earth. The feelings of Dr Jones, who had been one of the prime movers in the investigations during the previous ten days, can be imagined more readily than described. He pressed Eckersley to explain why he had produced the set of graphs, upon which he had himself relied so heavily in the Cabinet Room that morning. Eckersley renounced them, saying that they had applied to quite a different case when he had been trying to stretch his theory, as he himself admitted. The situation was not without its humour, since it had been these graphs that had first enabled Jones to convince Lindemann that *Knickebein* was a feasible proposition. Obviously someone was barking up the wrong tree. Jones could only hope it was not himself.

While Jones pondered on this unexpected development, the hunt for the beam signals continued. That evening an Anson aircraft again took off from Wyton, on the third of the investigation flights. The pilot was Flight-Lieutenant H.

E. Bufton, and the wireless operator was Corporal Mackie. On his return, Bufton reported:

1. There is a narrow beam approximately 400–500 yards wide, passing through a position one mile south of Spalding, having dots to the south and dashes to the north, on a bearing of 104°–284° True.

2. That the carrier frequency on the night of 21st-22nd June was 31·5 mc/s, modulated at 1150 c/s and similar to Lorenz in characteristics.

It would be difficult to overestimate the importance of this discovery. The early-warning radar stations which stood sentinel round the shores of Britain would enable RAF Fighter Command to handle roughly any bombing attacks launched during the hours of daylight. But there was no corresponding degree of success to be expected from the night defences. The hope had always been that the cloak of invisibility which had obscured the bombers from the ground and air defences would also hide the target from the bombers. Now it seemed that the Germans could operate effectively in the dark; the targets in Britain were naked and exposed.

It was vital that some means be found to counter this new threat. The German Air Force had not started large-scale night bombing attacks on England yet.[6] But they had the aircraft and the bases and it could only be a matter of time before they did.

It fell to Air Commodore O. G. Lywood, head of the Directorate of Signals at the Air Ministry, to initiate countermeasures. Lywood formed a special unit to counter the beams, an operation which rejoiced in the not inappropriate code-name of HEADACHE. The unit was No. 80 Wing, and its commanding officer was Wing-Commander E. B. Addison. 'Air Commodore Lywood and I had known each other for some time,' Addison later recalled. 'We had served together in Egypt and I was still out there when war broke out. He got me back from Egypt and I was posted to the Air

6. The *Knickebein* beams had been switched on merely to give crews practice in their use during the light probing attacks.

Ministry to work in his Directorate of Signals organization. I hated the job. One day he called me in and told me that a young fellow from scientific Intelligence called Jones had produced an extraordinary story of the Germans using a beam over this country to bomb London. It was known as *Knickebein*. He said: "I can't tell you how he came by this information, but he has. It looks extremely dangerous. What do you think we ought to do?" All I saw was a chance of getting away from this awful Air Ministry job, which just was not my cup of tea. I suggested that we needed a counter-measures organization – situated of course away from the Air Ministry.'

Addison could have had little idea of what he had talked himself into. Air Commodore Lywood acted on his suggestion, placed Addison in charge.

The third man in the team formed to fight the German beams was a young physicist who had joined the staff of the Telecommunications Research Establishment (TRE) at Swanage just a few weeks earlier – Dr Robert Cockburn. He and his small team of technicians took on the task of producing a radio jamming device capable of blotting out the *Knickebein* signals. This jammer would take time to come into service. In the meantime, it was essential to have ready some sort of jamming device, no matter how crude, if only to make the German aircrew realize that the British were tampering with their beams. If the enemy's confidence in the system could be reduced when the heavy attacks started, that at least was better than nothing.

Addison started by requisitioning a number of electro-diathermy sets – equipment used in hospitals to cauterize wounds – and had them modified to transmit a 'mush' of radio noise on the *Knickebein* frequencies in an effort to conceal the dots and dashes. Because of their low power, these sets were effective only within their immediate area. The diathermy jammers were in fact installed in selected police stations where the duty constable had instructions to switch them on when instructed to do so from the 80 Wing headquarters at Garston. Addison also secured some RAF Lorenz airfield beam-approach transmitters and had them modified to radiate a beam similar to that from the German *Knickebein* transmitters. The idea behind this was that the fake beam should be laid across the German beam, and the German aircraft would be pushed off course without their noticing it. Monitoring flights showed that the deviation caused was negligible, however, because there was insufficient power. But over short ranges the jamming signals did mask the 'continuous note' zone of the German beams, and a number of these sets went into service with 80 Wing.

Soon after the high-flying Anson aircraft had first stumbled across the *Knickebein* radio beam signals, a member of the RAF's listening service – perched uncomfortably with his equipment atop a 350-foot high radar tower – also managed to receive them. This was important: before any jamming action could be undertaken, it was vital to know which of their three frequencies the Germans were using; now it seemed that this could be done without having to keep aircraft permanently airborne. A little later it was found that the *Knickebein* signals transmitted by stations nearer to England – those on the northern coasts of France and Holland – could be received at ground level. By the end of July, listening teams were attached to many of the coastal radar stations.

While Addison strove to gather the equipment for some sort of makeshift jamming organization, there was another vital commodity he had to have – personnel. With the sort of highly technical war that was in the offing, Addison had no use for other people's misfits and throw-outs. He needed the very best material available. Fortunately, interest in the Wing's well-being extended from the Prime Minister downwards, and Addison was allowed freedom to choose those who served under him. It was even rumoured that 80 Wing accepted only the prettiest of WAAFs!

As the struggle to improvise measures to counter *Knickebein* went ahead, 80 Wing took control of another jamming commitment initiated some months earlier. During the last few months of peace, the BBC had made plans to prevent German bomber crews using broadcast transmissions to navigate their way about. From the declaration of war, all broadcasts went out simultaneously from a number of transmitters scattered throughout the country, all on the same frequency. It was thus impossible for any navigator to learn anything useful by tuning in his directional receiver and taking a bearing on these transmissions.

Apart from the possibility that enemy bombers might home on to BBC transmitters, there was another danger: they might home on to radio beacons planted for them by 'fifth columnists' – and of course there were the beacons the Germans had set up in their own territories. Early in 1940, GPO engineers worked out a means of countering enemy radio beacons, called the Masking Beacon, or 'Meacon'. The first 'Meacon' became operational on 24th July, at Flimwell near Tunbridge Wells.

Professor Lindemann explained how the system worked in a beautifully lucid paper sent to the Prime Minister on 10th August: the Germans, he said, had nearly eighty radio beacons in Germany, Norway and northern France, operating on the medium and long wave bands. Not more than twelve of these were used at any one time, the remainder being held in reserve; different groups were used on different days, with the enemy knowing which group was in action, and supposing the British not to know which call-sign corresponded to which station. If the enemy flew in such a way that two or more beacons were kept at some prearranged angle or angles to one another, he could easily pursue some prearranged course. The Prime Minister was told:

There are two ways of dealing with such beacons. The first is to jam them, i.e. to make so much disturbance in the ether that their signals cannot be received. If one compares them with lighthouses, it is like turning on the sunlight so that they would become invisible. This method is difficult because they operate on so many different wavelengths that we must produce very strong signals in each band to cover the lot. Furthermore, operators become very skilful in recognizing signals through the general glare. Though the signals are not such sharply defined points as lighthouses, it is clear that such a system of dots and dashes could not be recognized, even in bright sunlight, unless the illumination were excessively powerful. Further, each lighthouse has its own colour (wavelength) which has to be outmatched, so that the general glare must be produced over the whole spectrum, ranging from 30 metres to 1,800 metres. In order to cover this range eight very powerful stations would be required, but this leads us to another difficulty. If we had eight such stations, the Germans would soon get to know where they were and could use them as lighthouses to guide them to their targets. It is much easier to fly towards a beacon than to navigate away from it on back bearings.

In order to prevent this it would be essential to link our jamming stations in groups of three, making each group of three flash simultaneously. If this is done (though there is no exact optic analogy) the radio receiver cannot tell from whence the beam comes, so these could not be used as homing stations. On the other hand, this would imply the use of 3 × 8, i.e. 24 powerful stations, which would mean that all our home wireless had to be sacrificed for this purpose. By giving up the BBC and all other transmitters, this arrangement could possibly be made in four to six weeks. Even then we should have difficulty if the Germans chose to use, instead of their normal beacons, the super highpower stations normally used for wireless purposes in France and Holland.

This brings us to the second method, called 'Masking'. For this purpose, we require a number of small stations in England which pick up and repeat the German signals exactly in phase. If this is done, the wireless operator in the German machine cannot distinguish between the signals from his beacon and the echo signal from our station, and his direction-finding is completely set to nought. Since these echo stations are in exact phase with the ground stations it is impossible to home on them so that they cannot be used as a navigational aid by the enemy as a German station could. They are admittedly slightly more complicated to set up, but we have already six in action and a further nine will be operating within a week. Providing the Germans do not use more than twelve stations at a time we can mask them completely with these fifteen stations so this method of navigating will be nullified. All masking beacons are being provided as rapidly as possible and it is hoped in a few weeks to be able to cope with any possible German orchestra of beacons. Obviously, if we had eighty, we could deal with them if they turned on all their eighty beacons. On the other hand it is unlikely that they will use too many at a time as it would certainly confuse their own pilots very much. Thirty stations would probably suffice for anything the Germans are likely to do.

By 18th August, nine of these 'Meacons' were in operation. Two days later, the strange assortment of hastily erected HEADACHE stations – modified diathermy sets and Lorenz transmitters – was ready for the regular 'treatment' of the *Knickebein* transmissions. It was a close-run thing, for by now nine new *Knickebein* beam transmitters, strung out along the coasts of France, Holland and Norway, had been built to supplement the original three. On 28th August one hundred and sixty bombers of the Third Air Force (*Luftflotte 3*) delivered the first heavy night attack on England. The target was Liverpool. They returned in similar strength on each of the following three nights. The expected night onslaught had opened, but if the Germans had entertained great hopes for *Knickebein* they were to be disappointed.

On 7th September, the German air force shifted its night attacks to London, and from that evening until 13th November an average of 160 aircraft raided the capital every night, except for one occasion when the weather intervened. This German assault on London coincided almost exactly with the setting-up of the first of the high-powered jammers specially devised by Cockburn and his small team at Swanage to 'doctor' the *Knickebein* beams. The jammer bore the code-name 'Aspirin', and transmitted powerful Morse dashes on the German beam frequencies. These dashes were not synchronized with the German signals, but they were superimposed on them. The result was that when a German pilot entered the dash zone, he would turn in the required direction; but when he arrived in what should have been the central – 'steady note' – lane he continued to hear dashes and so tended to overshoot. When in the 'dot' zone, he would hear a mixture of dots and dashes, which would not resolve themselves into a clear note at all.

The 'Aspirins' were prescribed for the more important sites, replacing the less efficient jammers put together from modified diathermy and Lorenz transmitters; these latter were then moved to new sites to increase still further the area for which jamming cover was available.

During this jamming offensive, there was some controversy as to whether it might not be better to build a device to *bend* the German beams so that the enemy bombers could be pushed off course without the enemy realizing it. Technically a beam-bending device was quite feasible, but such an elegant countermeasure would have taken time to set up – and time was what Addison was short of now. If he was dealing with a danger threatening the very existence of Britain's great cities, he could not afford to waste time on subtlety. The margin of decision was admittedly a narrow one : in the case of the German radio beacons – Lord Cherwell's 'lighthouses' – the multiplicity of the frequencies used made subtle jamming deception by far the more economical course; but the present radio beams were radiated on only three frequencies, and cruder jamming was the immediate remedy.

Perhaps it began as a wartime 'plant' by British Intelligence, with the aim of weakening German confidence in their beams, but the fact is that the deliberate 'bending of the beams' is still widely believed to have happened during the war, and even then the story was quite widespread. This sometimes proved to be a source of embarrassment to 80 Wing: 'Whenever anything unusual happened, people thought it was us,' Addison recalls. 'At the time we worked in such secrecy that when these funny ideas got around we had no means of correcting them – even had we wanted to. On one occasion a German aircraft unloaded its bombs in the castle grounds at Windsor. The next morning, the Comptroller of the King's Household rang me; he was very cross, and wanted to know why we had bent the beams over Windsor – His Majesty might have been killed. It was the usual case of a lost German getting rid of his bombs – and we got the credit or the blame, depending on where they fell.'

On the same subject, Dr Cockburn says: 'The myth has been established that we bent the beams. In fact we didn't. I did rig up a system using a receiver at Worth Travers, near Swanage, and a transmitter at Beacon Hill, near Salisbury. I was going to pick up the modulation of the *Knickebein*, retransmit it, and thus push the beam over. In other words, my transmitter would have produced a beam similar to the German ground station but pointing to where I wanted it to. It was all very nice, but it didn't happen. By the time the system was ready, the other jamming methods were in full swing and we could not spare the time or the effort to bring out a new system to supplement the old. So the deliberate bending of the German beams, which I had worked out, never happened.' When 80 Wing took over the Beacon Hill station they employed it to transmit unsynchronized dashes, just like their other 'Aspirin' jammers.

The bombing of Dublin, some six months later, has often been quoted as proof that the RAF did bend the German beams; indeed, the impression appears to be widespread that 80 Wing deliberately caused a force of German aircraft to bomb the neutral Irish capital. What really happened was that early on the morning of the last day in May 1941 about ninety German aircraft raided Bristol and Liverpool, and at the same time bombs fell in the North Strand and Phoenix Park areas of Dublin: they demolished twenty houses and damaged fifty-five; twenty-eight people were killed and eighty-seven were injured. In fact, the Eire defence department had estimated the strength of the attacking force as just one aircraft! As a people, the Irish never have been noted for an ability to play down a story. In all probability this was just a German crew losing its way: that they had drifted so far to the west may well have been attributable to the wind, which blew that morning at thirty knots from the east.

But if the *Knickebein* beams were not bent deliberately, they were certainly bent accidentally on occasions. The 'Aspirin's' dashes were not radiated in synchronism with the beams' dashes, it will be recalled. But they were not, on the other hand, deliberately radiated *out* of synchronism. The result was a continuous drift in and out of synchronism, rather as two people walking together with different lengths of pace will drift into and out of step. So there were times when the 'Aspirin' dashes synchronized accidentally with the German dashes, and other occasions when the British transmitters were so far off tune – or the German signals so weak – that the enemy aircraft tuned their receivers to the fake dashes alone. In either case, the aircraft would continue turning in the hope of steering towards the zone of 'dots', and the result was a bending of its flight path. This does appear to have happened, and it certainly became a current belief within the German air force. One captured airman even related how he had heard of pilots who had involuntarily described a full circle; and on other occasions, German crews discovered that they were a long way from where they thought they should have been, and they blamed their error on the beams.

In October 1940, Addison was made a CBE and promoted to Group Captain. No. 80 Wing now comprised twenty officers and two hundred men, and operated fifteen 'Aspirin' sites to interfere with the *Knickebein* beams. What had been achieved? In theory, the beams could have marked out a 'square' with sides 300 yards long in the sky over London – about the area of one of the larger blocks of Ministry buildings in Whitehall. Given the undisturbed use of the beams, each raiding German bomber could have planted its bomb-load within this area; if just one-quarter of the bombs from an attacking force of 160 aircraft had been dropped with such accuracy, there would have been a saturation of one bomb every 17 yards throughout the target area.

During their night offensive against London, the German air force mounted sixty-seven attacks on this scale. That no such saturation occurred during any of the raids demonstrates the effectiveness of the HEADACHE countermeasures.

By using *Knickebein* in small practice operations, before they were in a position to back it with large-scale night attacks, the Germans had compromised its secrets. When their air force really did need the beams, 80 Wing already had the knowledge and the means to wreck them.

The campaign against the *Knickebein* beams had been fought on the highest technical level, by a very few. The men of 80 Wing had not experienced great personal danger; in general they had suffered nothing worse than boredom, discomfort and long hours. But their contribution towards the saving of London was no less than that of the fighter pilots who defended the capital more conspicuously by day. The secret defeat of the 'crooked leg' beams was also a great personal triumph for Dr R. V. Jones and scientific Intelligence: it ensured that the next time the watch-dog barked, people would listen and comply. But there was no time for anyone to rest on his laurels, for the battle against the beams was far from over. The Germans still had their 'X-beams'.

In the middle of August 1940, the German air force had begun the all-out aerial bombardment of England. On the evening of 13th, 'Eagle Day', when the first of the hard-fought daylight battles was over, a force of about twenty German aircraft had bombed a factory near Birmingham which was tooling up for the production of Spitfires. What was unusual about this attack was the inordinately high degree of concentration for a night raid – eleven bombs had actually hit the factory buildings. Had anyone been in a position to notice, he would have seen that a colourful emblem was stencilled on each aircraft's nose – a Viking ship. Each aircraft carried the complex *X-Gerät* developed by Hans Plendl before the war began. This was the device which picked up the three 'cross-beams' as they intersected the main approach beam, and timed the release of the bombs automatically against a special clock. All the aircraft belonged to the special beam-flying squadron, *Kampfgruppe 100*.

After this start, *KGr100* aircraft had taken part in a series of small raids which had achieved little, although they did provide useful training for their crews. The squadron next operated in force in conjunction with the rest of the German air force in a series of heavy attacks on Liverpool at the end of August. These small beginnings were enough for the British Intelligence services to open a file on this new system. The RAF code-named the system 'Ruffian'. Half-way through August, their monitoring service picked up some unexplained signals on 74 megacycles. As this frequency was outside the range of any receiver known to be used by the German air force, the report was treated with some reserve. But by the end of the month there was ample confirmation: not only had monitors noted more signals on nearby frequencies but ground direction-finders had located sources of them in the Calais and Cherbourg regions. Although the signals differed from *Knickebein* in their frequency, the pitch of their tone and the rate of their 'keying', they were sufficiently similar for them to be identified as an aid to navigation.

During the following month, Jones added more and more pieces to his jigsaw picture of the device, although no actual equipment had fallen into his hands. The danger it represented became very evident. At the end of the third week in September, the Prime Minister was informed:

It appears that the Germans are making great efforts to improve the accuracy of their night bombing. A number of new beams on a shorter wavelength than before have appeared . . . One *Kampfgeschwader*,[7] *KG100*, consisting of about forty machines, has been equipped with special new apparatus to exploit these beams with which apparently accuracies of the order of 20 yards are expected.

With the technique they seem to be developing such a

7. The unit was *Kampfgruppe 100*, not a *Kampfgeschwader* (bomber wing). In his report to Churchill, quoted above, Professor Lindemann appears to have confused the two units as he gave the *Gruppe* (squadron) strength of 40 aircraft correctly.

result does not seem impossible. We know the exact location of the sources of the beams in question. The parent beam is on the very tip of the Cherbourg peninsula; the cross-beams are in the Calais region. They will probably not reach much beyond London. Apart from attacks on the machines using the beams, our possible lines of defence would be:

1. to try to destroy the specially fitted *KG100* machines which are stationed at Vannes; home station Luneburg and reserve station at Köthen;
2. to try to destroy the beam stations
    (*a*) by bombing, which would be very difficult since they are almost invisible targets
    (*b*) by a special [i.e. commando] operation;
3. to employ radio countermeasures.

The estimated accuracy of twenty yards was based on the degree of accuracy to which the Germans were known to have set up their beams, and not on any technical assessment of the capabilities of the *X-Gerät*. In fact the mean error was of the order of 120 yards, which was still a very low figure.[8] Of the alternative 'medicines' proposed in the report to Churchill, the least difficult was the employment of radio countermeasures. But even this was not easy: the British had no 'off the shelf' transmitter that would radiate on a frequency around 70 megacycles, so Dr Cockburn modified an Army radar set to jam the 'X-beams', and gave it the code-name 'Bromide'. His section worked at high pressure to produce the numbers of these jammers necessary to provide a minimum of cover. The immediate plan was to install the 'Bromides' between Cherbourg – the

8. When calculating the angular alignment of the X-beams the Germans almost certainly made use of the seven-figure trigonometrical tables worked out by Dr Hermann Brandenburg of Göttingen university. When he had completed them some time before the war, Brandenburg had been so anxious to ensure that they were correct that he had offered the sum of three shillings for each mistake that could be discovered. Dr Cromrie, one of the British scientists now engaged on back-plotting the X-beams to their sources, had taken about £20 from Brandenburg in this way.

source of the parent, or approach, beams – and the Midland towns, Manchester and London.

While the 'Bromide' transmitters were being built, the Germans were able to use *X-Gerät* unhindered. By the end of September, *KGr100* had taken part in some forty attacks, half of them on London. During this period, the squadron operated as an independent force, visiting targets alone and attempting precision attacks by means of their beams. These small-scale attacks all bore the same fingerprints, which R. V. Jones was able to trace back to the German specialist squadron: there was great accuracy along a line running from Cherbourg, slightly lower accuracy in range, and no bombs heavier than 250 kilograms were dropped.

Early in October, the squadron started dropping incendiary bombs. This was on the face of it a strange development, as these stick-shaped weapons could not be accurately aimed, which seemed to nullify the main advantage of the X-beam system. There seemed only one reasonable explanation: *KGr100* was practising to lead the now '*de-Knickebeined*' air force to targets.

On 24th October, Professor Lindemann accordingly advised Mr Churchill:

There is some reason to believe that the method adopted is to send a few *KG100* aircraft fitted with special devices to assist in blind bombing on these expeditions in order to start fires on the target which any subsequent machines without special apparatus can use.

The note accurately predicted the course the Germans were to adopt three weeks later, which was to result almost immediately in the catastrophe at Coventry. This was the advent of the German air force 'pathfinder' techniques.

In the meantime, fate was about to play into the hands of British Intelligence in one of the more inept episodes of the secret war. Early on the morning of 6th November, a raiding Heinkel bomber suffered a compass failure over England. Using radio bearings from the beacon at Saint-

Malo, the crew turned for home. They flew right over the beacon and continued until they felt themselves safely over southern Brittany. The pilot descended, but as he broke cloud he saw that he was still over the sea. This had to be the Bay of Biscay, so he turned back to the beacon, and land. By this time his fuel was almost exhausted. When a coastline came into view he attempted to put the bomber down on the beach; he misjudged his approach, however, and in the resulting crash one of the crew was killed and two were injured.

Those who survived clambered up the shingle beach to find themselves surrounded by soldiers in khaki. The treacherous beacon in which they had placed their trust was in fact one of 80 Wing's 'Meacons'. And what they had thought to be the south-western coast of Brittany was the beach at West Bay, near Bridport. Some of the British soldiers waded out and secured a rope round the wreckage, and all would have been well had a Royal Naval vessel not arrived. The ship's captain pointed out that since the bomber was in the sea, its salvage was technically a naval matter. After some wrangling the Army grudgingly agreed. The sailors took the line aboard their ship and solemnly towed the aircraft out into deeper water, preparatory to lifting it out. Unfortunately the rope broke, and the Heinkel sank to the bottom. When dawn broke, the top of the wrecked bomber could be seen above the waves, looking rather like a stranded whale. As the tide receded, the markings gradually came into view: painted on the rear fuselage in characters nearly four feet high was the cypher, 6N+BH; on its nose was the emblem of a Viking warship. As the RAF's monitoring service had discovered, 6N was the squadron call-sign of *KGr100*.

Professor Lindemann was understandably bitter when he learned what had happened. A week later, he wrote to the Prime Minister:

The *KG100* squadron is the only one known to be fitted with the special apparatus with which the Germans hope to do accurate night bombing using their very fine beams.

As it is important to discover as much as possible about this apparatus and its mode of working, it is a very great pity that inter-Service squabbles resulted in the loss of this machine, which is the first of its kind to come within our grasp.

All was not lost. British technicians managed to remove the two invaluable X-beam receivers from the aircraft's waterlogged hulk. Somewhat the worse for their prolonged immersion, the pieces went off for examination. Any last doubts still nursed about the advanced nature of German radio beam technique were dispelled by the dates on some of the receivers' inspection-stamps. They went back to 1938.

In the second week of November, the German air force shifted the burden of its attack from London to the Midlands cities, and *KGr100* was to lead the way for the bomber force in the first of these new raids. The target was Coventry. On the afternoon of 14th November, the Boulogne headquarters of the German 6th Air Signals Company – the unit which operated the X-beam transmitters – received its orders on the night's target from the squadron's headquarters at Vannes. These were passed on to transmitters *Elbe*, *Oder* and *Rhein* nearby, and to *Weser*, the 'approach-beam' transmitter, on the Cherbourg peninsula. The approach-beam crossed the English coast near Christchurch, ran just to the east of Salisbury and Swindon, and passed over Leamington and Coventry. Shortly before the latter city the three 'cross-beams' intersected it.

Soon after dark the Heinkel 111's of *KGr100* took off from Vannes. As a small concession to the British defences, they flew to one side or other of the main approach beam, as the beam itself might be patrolled by night-fighters. By 8 p.m. the leading aircraft had crossed the line of the Thames near Bampton. Six minutes later they flew through the first cross-beam, and funnelled into the approach beam and set course for Coventry.

There were on this night only four 'Bromide' stations in operation, ready to deal with these new X-beams. One, at

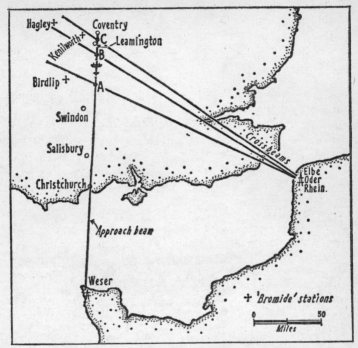

The Layout of the X-beams Over Coventry
on the Night of 14th November 1940

Point A – the first cross-beam (*Rhein*): the aircraft
closes on to the approach beam (*Weser*). The distance
from A to B is 30 kilometres. Point B – the second
cross beam (*Oder*): aircraft observer presses button
to start the bombing clock. The distance from B to C
is 15 kilometres.
Point C – the third cross-beam (*Elbe*): aircraft observer presses button to stop first hand of bombing
clock; second hand moves round towards the first. The
distance from C to Coventry is 5 kilometres.
Target Coventry: hands on bombing clock overlap,
closing pair of electrical contacts to release bombs
automatically.

Kenilworth, was almost directly beneath the bomber's run in. But the jammers caused the Germans little trouble: the system had been hastily conceived at a time when too little had been known about the *X-Gerät,* and as a result although the jammers probably emitted on the correct frequency the note they put out was modulated at 1,150 cycles instead of 2,000 cycles. The difference – it was the difference between a whistle and a shriek – was just perceptible to the human ear; but the filter circuits in the German receivers were sensitive enough to pick the beam signals out of the jamming with ease. Only if all the jamming had been of exactly the right type would the four 'Bromides' have sufficed to render the X-beams unusable. On this night it was not.

The Heinkels continued on their northerly course undisturbed, and at eleven minutes past eight, three miles south of Leamington, they flew through the second cross-beam. In each aircraft the observer started his special automatic bomb-release clock. Two and a half minutes later, less than a mile east of Bagington, they flew through the third and final cross-beam: the second pointer on each clock started moving round, catching up on the first. The die was cast. In fifty seconds, the two pointers were overlapping and the pairs of electrical contacts were closed; the incendiary canisters were released. It was a quarter-past eight.

The fires started by *KGr100* in Coventry guided the bombers in from all directions. One stream came in over the Wash, another over the Isle of Wight and yet a third over Brighton. The night was clear and in the moonlight the approaching crews were able to see every detail of the burning city. Altogether 449 bombers hit Coventry during the ten hours of the attack. Between them they dropped 56 tons of incendiaries, 394 tons of high-explosive bombs and 127 parachute mines. Each of the German bomber units had its own specific target to search out and destroy: *I/LG1* was briefed to attack the works of the Standard Motor Company and the Coventry Radiator and Press Company; *II/KG27* the Alvis aero-engine works; *I/KG51* the British Piston-Ring Company; *II/KG55* the Daimler works; and

KGr606 the gas-holders in Hill Street. Most of these objectives were heavily hit, and all production ceased in consequence of the debilitating damage to the city centre. Nearly four hundred people died and a further eight hundred were seriously injured as a result of the attack. It was a striking demonstration of what the German air force could do using a 'pathfinder' force guided by radio beams.

The deficiency of the 'Bromide' jammers was deduced from the captured *X-Gerät* receiver. Within a few days of the Coventry catastrophe, the old 'Bromide' sets had all been modified to transmit the correct jamming note, and new sets were emerging from Cockburn's workshops at an encouraging rate. It was a much more effective 80 Wing that went into action to counter *KGr100*'s next 'pathfinder' marking effort. This time German squadron's target was Birmingham, on 19th November: again it was a clear night, but thanks to the jamming from the newly-modified 'Bromides' *KGr100* found considerable difficulty in hitting the city and only a few small scattered fires were started to the south of it. When the main force aircraft arrived, they wandered aimlessly for some time before finally unloading their cargoes over a wide area. The attack was a failure, as was a similar one the following night. Nevertheless, the German 'pathfinder' squadron continued to ride the crest of the wave: it was the most distinguished unit in the German bomber force. Even during December 1940 there were still cities in England without adequate 'Bromide' cover, although their number was dwindling, and against these the German squadron achieved some successes: London, Southampton, and Sheffield suffered especially heavily from attacks for which *KGr100* had marked the targets. But the *X-Gerät*, the squadron's own 'magic box', was inexorably approaching its premature eclipse.

While the British jamming offensive was gaining in momentum, other measures were being taken to exploit the great weakness of the German 'pathfinder' tactics. The main body of the German raiding forces released their bombloads on fires kindled by the 'pathfinder' aircraft, so why not light fires in the open countryside for them to bomb instead? The

job of setting up these decoy fires – known as 'Starfish' – fell
to a section headed by Colonel J. Turner, one-time head of
the RAF works department. The decoys had to look plausibly like cities under attack, and the timing of the blaze was
critical: it had to start just at the right time to catch the main
attack, and the bombers should ideally have to fly over it
before reaching the real target. The first 'Starfish' were ignited on the night of 2nd December 1940, during an attack
on Bristol. The two sites collected a total of 66 high-explosive bombs. From then on, these decoys became a regular
feature of the British passive defences. It was found that if
the fire services could deal promptly with accurately placed
German marker fires, a decoy on the approach might well
draw a high proportion of the bombs intended for the town.
Crews in the leading aircraft invariably reported by radio
on the success of their marking; if they had done so, the
aircraft following were quite content to bomb any reasonable conflagration in sight.

By the middle of January 1941, every major target in the
southern half of England had a 'Bromide' jammer to cover
its approach line from Cherbourg. There were even enough
jamming transmitters to blot out some of the cross-beams. In
March, a captured crew from *KGr100* confirmed that the
interference to the beams had become progressively worse,
ever since the beginning of November. The jamming of the
approach beams had become very serious by the end of the
year, and early in 1941 No. 80 Wing noted the first unmistakable signs that the German Air Force was losing confidence in the *X-Gerät* as a device for 'pathfinder' marking:
twice during March *KGr100* had not appeared at the target
until after the attack had opened; and on one occasion they
dropped parachute mines – weapons which could not be
aimed with any accuracy at all. Even when the beams were
used, they underwent complex frequency changes while the
attack was actually in progress. During May, the 80 Wing
monitoring stations heard the X-beams on all but three
nights. The 'pathfinder' squadron operated on fifteen
occasions, but on five of these the radar plots showed

that they did not use their beams. The 80 Wing diarist recorded:

> It would thus appear that while the enemy still has considerable faith in the skill of *KG100*, his regard for 'Ruffian' [*X-Gerät*] is steadily declining. The combined evidence of increased complication in an attempt to avoid our countermeasures, and the failure to make use of this system in its present state on 33⅓ per cent of operations is too strong for this view to be ignored.

Several times during the war, British Intelligence arrived at the right answer by what proved to be a wrong process of reasoning. The detection of the Germans' third radio beam system was an example of this. At the end of June 1940, Dr R. V. Jones had received a report that '*Wotan*' transmitters were being erected near Cherbourg and Brest. What on earth could *Wotan* be? The text of the secret report implied that the device was some form of navigational aid; but that was all there was to go on. Fortified by a modicum of mythological research, Jones felt able to make a shrewd guess at the nature of *Wotan*: in the past the Germans had sometimes used code-names which, while very clever, had betrayed the nature of the device they were intending to conceal. Wotan – alias Odin, the leading deity in the kingdom of Nordic mythology – had only one eye. Was it beyond the bounds of possibility that by using the range-finding device described in the Oslo Report to measure the range, and one of the newly-discovered *Knickebein* beams for direction, an aircraft would be able to fix its position using only one beam instead of two?

In November 1940, the month of the Coventry raid, the RAF monitoring service first noticed some unusual signals on frequencies around 40 megacycles. Jones allocated the system the code-name 'Benito'. (He says: 'We reckoned that since Mussolini was at the one-eyed end of the Axis, it was appropriate to call the single-beam system "Benito".') At

first the new signals defied explanation: the very high rate
of transmission of the directional signals made it impossible
for the listeners in England to sort out what was going on.
Only after they had been examined on a cathode-ray tube
did their nature become clear. Thanks to the information
given in the year-old Oslo Report, there was no similar
difficulty in working out how the ranging system operated.
As Jones noted:

> There can now be little doubt that, whatever his
> motives, the anonymous source was giving a scrupulous
> account of his own wide knowledge and, for his timely
> warning of *Benito* alone, we are his debtors.

Like its predecessor the *X-Gerät*, the new system had
been developed by the German Dr Hans Plendl; the German
air force in fact called the new system *Y-Gerät*. It was a
complex system: to align the aircraft on the target, the
ground transmitter radiated a rather complicated beam
made up of 180 direction signals per minute. This was too
fast for human interpretation, and the aircraft carried a
special electronic analyser to work out its position relative
to the beam. The gap which followed each pair of direction
signals served to lock the analyser to the beam. This complex beam was the one which caused the British monitors
the initial difficulty.

In order to measure the aircraft's range *along* the beam,
the ground station transmitted additional signals; the aircraft picked these up and re-radiated them. The ground
station was thus able to compute the aircraft's range at much
greater distance than conventional radar; when they computed that the aircraft was at the bomb release point, they
radioed instructions to the crew to release the bombs. Both
the range and the bearing signals were radiated on frequencies between 42 and 48 megacycles.

Since it used only one ground station, *Y-Gerät* was a more
flexible system than either of its predecessors; and it was
even more accurate than *X-Gerät*. General Martini, the
wartime head of the German Air Force signals service, has

told of how he once tried to explain the working of *Y-Gerät*
to Hermann Göring. The Reichsmarschall is said to have
listened for about two hours, then asked some questions
which showed clearly that he was none the wiser. Göring, a
World War I fighter ace, had little time for such gadgetry :
wars were to be fought by brave men with guns, not like
this. He is said to have commented on another occasion,
'Radio aids contain boxes with coils, and I don't like boxes
with coils.' It is hard not to feel for him.

Plendl's new system was ready by the end of 1940. The
Y-Gerät was fitted to the Heinkel 111's of the third squadron of *KG26* based at Poix, near Amiens, and the squadron
began operational trials using ground transmitters at Poix,
Cherbourg and Cassel in France. Much later, Dr R. V. Jones
learned that the full designation of *Y-Gerät* was in fact
Wotan II; and that of *X-Gerät* was *Wotan I*, and this latter
system was very much a multi-beam system. The truth was
that the god's one eye had had nothing at all to do with the
Y-Gerät's other code-name! Jones now admits that had he
known this from the start, he might have taken much longer
to guess the real method employed by the new system.

As *Y-Gerät* employed two separate transmissions to fix
the aircraft's bearing and range from the beacon, Dr Cockburn had to work on a separate jamming method for each.
For the first time there was no great rush to get jammers
into operation – the system was still at the working-up stage.
There was even time to be subtle about the countermeasures.

Cockburn's jammer – code-named 'Domino' – employed
a receiver at Highgate and the BBC's dormant television
transmitter at Alexandra Palace, in North London. The
receiver was used to pick up the ranging signal 'echoed'
from the German bomber's transmitter; the signal was passed to Alexandra Palace, where the powerful transmitter re-
radiated the 'echo' on the frequency of the German ground
station. The effect on *Y-Gerät* was catastrophic, and the
system of range measurement was completely ruined. The
first such 'Domino' jammer went on the air in the middle
of February 1941, and a second was ready before the end of

the month on Beacon Hill near Salisbury. As far as Jones,
Cockburn and Addison were concerned, the jamming of
Y-Gerät was a complete success story. The device never
had a chance to get going before it was stopped in its tracks.
'Benito's' one eye had been gouged.

It was obvious that the 'Domino' transmitters were a
great embarrassment to the bomber crews. On 9th March,
the Y-beam signals changed frequency in the middle of an
operation in an attempt – unsuccessful – to shrug off the
jamming. Two nights later, a force of bombers devoted a
special attack to the Beacon Hill jammer, and one scored a
near miss. On the following night the jammer was off the air
when *KG26*'s third squadron operated. Some of the aircraft
used the Cassel transmitter, but the Alexandra Palace jammer was covering that one and none of the crews received
the order to release their bombs. Those aircraft which were
using the Beaumont transmitter – for which there was no
'Domino' cover – all carried out highly accurate attacks.
But on the next night the Beacon Hill station was back on the
air and once again the jamming was 100 per cent. Of the 89
Y-Gerät sorties over England during the first two weeks of
March 1941, only eighteen resulted in aircraft receiving
instructions to release their bombs.

On the night of 3rd May, the Y-beams squadron suffered
an even worse disaster: three of its Heinkels were lost over
England, and in each case the special *Y-Gerät* equipment
was carefully removed from the wrecked aircraft and sent
off to Farnborough for investigation. The examination
showed that the electronic bearing-analyser was vulnerable
in the extreme to jamming. As Cockburn put it later: 'Un-
locking the Y-beam was a piece of cake: they had fallen
into the trap of making things automatic, and when you
make things automatic they are more vulnerable. All one
had to do was to radiate a continuous note on the beam's
frequency. This filled in the gap between the signals, un-
locked the beam analyser and sent the whole thing haywire.'

The new jammer first went on the air on 27th May.[9] A
simple additional circuit – a DC restorer – in the *Y-Gerät*

9. It was code-named 'Benjamin'.

receiver in the aircraft would have filtered off the interference from the jammer, but the Germans did not get around to it. For the Germans, time was running out: from the middle of May, first in a trickle and then in a flood, the Air Force units moved to the east in preparation for the invasion of Russia. The bombers were scheduled to return to France six weeks after the opening of the offensive – a conservative estimate of how long the campaign was expected to last.

Group Captain Addison could not bank on the respite's being more than a brief one. During the summer and autumn of 1941, the build-up of 80 Wing continued, but it lacked the high priority it had enjoyed before. Gradually the makeshift jammers of the previous autumn were replaced with equipment properly designed for the job. The intensity of the Blitz gave way to the boredom of waiting for an enemy who rarely came.

The highlights of this period were when German aircraft were drawn off course by the 'Meacons': sometimes the enemy's confusion was so great that the aircraft landed in England by mistake. Late in July, the RAF seduced and secured an intact Junkers 88 in this way, and on 21st October the 'Meacons' proved their worth again: a German aircraft, a Dornier 217, was returning to its base at Evreux after a reconnaissance sortie over the Atlantic, when its crew encountered winds stronger than those forecast. Unknown to them, their aircraft had drifted well to the north of the expected track, and when the pilot did not cross the west coast of France at the time expected, he turned northwards hoping to get a fix off the south coast of England. Eventually he made a landfall over Pembrokeshire – which he mistook for Cornwall. He accordingly turned south and crossed the north Devonshire coast, which the navigator mistook for Brittany. So far their misfortunes were the result of bad luck, but now the RAF took a hand: No. 80 Wing was operating 'Meacons' at Templecombe and Newbury to cover the German beacons at Piampol and Evreux respectively. The Dornier meandered almost the whole length of southern England, its crew growing increasingly puzzled when the maps they carried did not appear to agree

Liverpool
Manchester

Birmingham

LONDON

Mt. Violette

Greny

Beaumont Hague

Sortosville-en-
Beaumont

Morlaix

Mt. Pincon

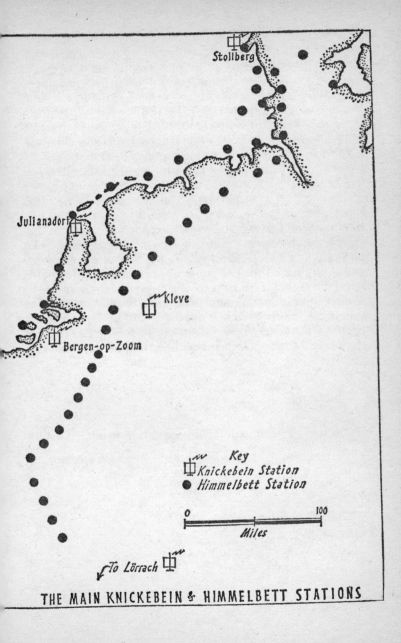

Stollberg

Julianadorf

Kleve

Bergen-op-Zoom

Key
Knickebein Station
Himmelbett Station

0   100
Miles

To Lörrach

THE MAIN KNICKEBEIN & HIMMELBETT STATIONS

with any of the 'France' they were passing over. Eventually, the pilot turned south, quite lost, and to his amazement came to yet another coastline. By now he had exhausted almost all his fuel; he was forced to land at the first airfield he came to, which turned out to be Lydd in Kent. The intact Dornier 217 thus presented to the RAF was the first example of this new bomber to be examined by British Intelligence.

Incidents like these occasionally punctuated the calm over England. But No. 80 Wing had won its first big battle, and the men who had done so much to make the unit strong were moved on to other tasks. The first round in the radio countermeasures campaign had gone squarely to the British. It would, however, be unwise to conclude from this that German technicians could not have modified their radio aids to operate more effectively in the face of the jamming, had they directed their attentions to that end. The truth was that before they could do so, the night bombing offensive against Britain had ground to a halt. If the battle of the beams had been a rather one-sided affair, the other encounters in this radio war were not to follow this pattern. As we shall see, the Germans could do much better than this.

2

The Instruments

I have done my best during the past few years to make our air force the largest and most powerful in the world. The creation of the Greater German Reich has been made possible largely by the strength and constant readiness of the air force. Born of the spirit of the German airmen in the First World War, inspired by its faith in our Führer and Commander-in-Chief – thus stands the German Air Force today, ready to carry out every command of the Führer with lightning speed and undreamed-of might.

– Order of the Day from HERMANN GÖRING to the
German Air Force, August 1939

Radar, like most of the major technological advances during the twentieth century, did not result from a sudden and inspired line of thought pushed to the point of fulfilment by one inventor. As with other great innovations, the basic idea preceded the invention by several decades, and it was only when certain special means had been developed that its realization became practicable. Again, as with the other great inventions of this century, once the background work was complete development proceeded independently in several nations simultaneously.

The development of the aeroplane and the turbo-jet engine are good cases in point: Sir George Cayley had laid down the important scientific principles necessary for powered flight as early as 1857. He failed to build the aeroplane as we now know it merely because he had no means of producing the necessary power without incurring a crippling weight penalty. When this became technically possible, the Wright brothers and Langley in America, and

Ader in France were all working along broadly similar lines. In the case of the turbo-jet engine, Mélikoff had designed a 'helicopter' as early as 1877, whose rotor was to be rotated by a gas turbine, 'consisting of eight curved chambers, into each of which charges of the vapour of ether mixed with air are to be successively exploded by an electric spark, and the charge allowed to expand in doing work'. This line of thought had come to nothing at the time because the piston engine was far more efficient at low speeds than the turbo-jet; moreover, the metals able to withstand the extremely high temperatures generated in gas turbines did not exist for another half century. When they did, the device reached fulfilment in the late 1930s, quite independently, in both Britain and Germany.

This pattern of 'pre-invention' holds good for radar as well: for it was on 30th April 1904 that the Royal German Patent Office granted a patent, to cover the basic radar idea, to a young German inventor, Christian Hülsmeyer. His device comprised a wireless transmitter and receiver mounted side by side, and so arranged

that waves projected from the transmitter can only actuate the receiver by being reflected from some metallic body, which at sea would presumably be another ship.

Hülsmeyer called his invention the 'Telemobiloscope'. It rang a bell whenever the receiver picked up echo signals. Some sources quote the device as having a range of several hundred yards. This may well have been so if it was mounted on flat ground with no reflective surfaces nearby, but it would not have worked on a ship or anywhere else where it might have been of use: the device depended on the transmissions being thrown forward in a narrow pencil beam; but nobody really understood the nature of radio waves in 1904, and the efficient beaming of radio signals was not to be achieved for another quarter of a century. With his crude aerials, Hülsmeyer could not possibly have determined the direction from which the echo signals came, and he must have picked up a lot of unwanted echoes from objects nearby and even behind the set. On a ship, reflections from the bridge, the funnels and even the lifeboats would all have caused the bell to ring; and there were other problems. The strength of the signals which bounce off metallic objects is always very low. In a modern radar set they are amplified by a factor of several million before they are displayed on the cathode-ray tube. But in 1904 there were no means of amplifying radio waves, so only echoes from objects very close to the 'Telemobiloscope' would have been noticed. Hülsmeyer took out patents for his device in a number of countries, but nobody seems to have bought one.

The advances in electronic technology necessary for the idea to work were realized during the next twenty-five years. The cathode-ray tube already existed in a primitive form: Professor Ferdinand Braun had built one in 1897, and to this day the device is known in Germany as the 'Braun tube'. In England, Professor Ambrose Fleming built the simplest form of radio valve, the diode, in 1904, and three years later an American scientist, Lee de Forest, added a third electrode to this diode to produce a 'triode'. The triode enabled radio signals to be amplified for the first time.

Twenty years after Hülsmeyer's invention, Gregory Breit and Merle Tuve, working together in America, were the first to use pulsed radio transmissions to determine the range of a distant object. Their work was actually directed at finding the altitude of the layer of ionized gas surrounding the earth, by measuring the time difference between the transmission of the signal upwards, and the return of its echo. They found that the altitude of this layer, which obviously reflected radio waves extremely well, was seventy miles. Four years after these experiments, in 1929, a Japanese professor, Hidetsugu Yagi, published an important paper on the transmission and reception of radio waves using directional aerials: for the first time it was possible to send out radio signals in a reasonably fine beam.

Thus by the beginning of the 1930s, the complete technological background for the development of radar existed. In 1935 British scientists began their series of now well-known experiments, which were to yield an efficient chain

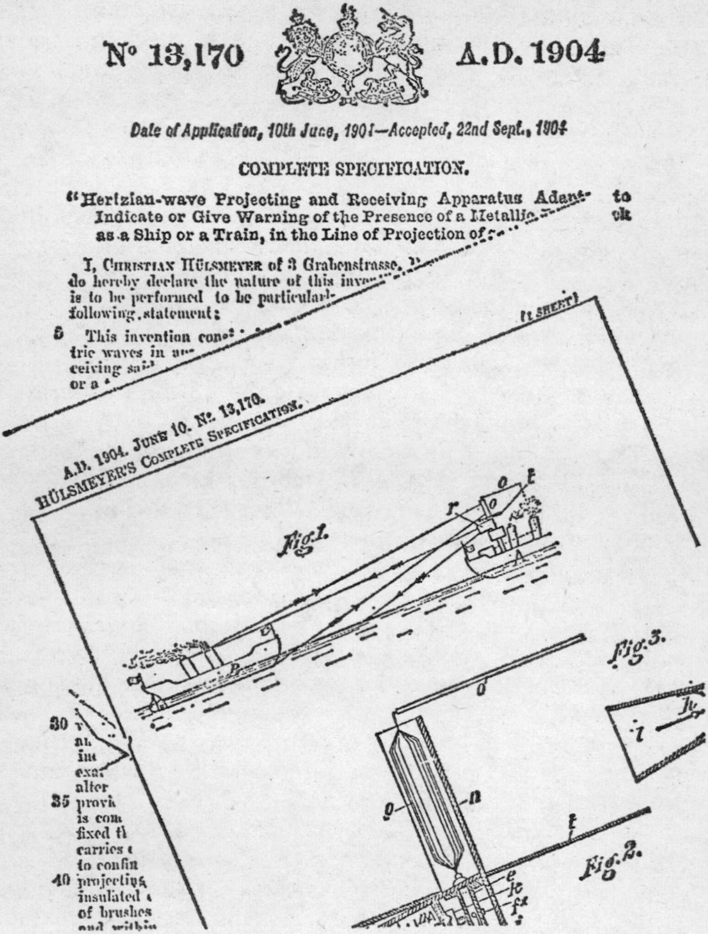

N° 13,170   A.D. 1904

Date of Application, 10th June, 1904—Accepted, 22nd Sept., 1904

COMPLETE SPECIFICATION.

"Hertzian-wave Projecting and Receiving Apparatus Adapt~ to
Indicate or Give Warning of the Presence of a Metallic ~~~~ ck
as a Ship or a Train, in the Line of Projection of ~~~"

I, CHRISTIAN HÜLSMEYER of 3 Grabenstrasse, ~~~~~~
do hereby declare the nature of this inve~~~~~
is to be performed to be particular!~
following statement:

5   This invention conc~~~
tric waves in u~~
ceiving sai~
or a ~

(1 SHEET)

A.D. 1904. JUNE 10. N° 13,170.
HÜLSMEYER'S COMPLETE SPECIFICATION.

Fig.1.

Fig.3.

Fig.2.

30   v~
   an~
   in~
   exa~
   alter
35   provi~
   is com~
   fixed t~
   carries ~
   to consi~
40   projectin~
   insulated ~
   of brushes
   and withi~

Hülsmeyer's British Patent for the
'Telemobiloscope'

of radar stations in time for the Battle of Britain five years later. But no nation holds a monopoly in good ideas. While radar was slowly being forged into a weapon of war in Great Britain, German scientists were also bringing it to fruition, along an almost identical path of development.

Dr Rudolph Kühnold, the head of the German Navy's signals research department, had been working on devices to detect underwater objects by bouncing sound waves off them – the apparatus now known as Sonar. Early in 1933 it occurred to him that what worked under water with sound waves might also work above the surface using radio waves. That summer he started experimenting, in complete ignorance of what Hülsmeyer had tried twenty-nine years before. His first device, which worked on a frequency of 2,000 megacycles, proved a failure because he was unable to procure any valves capable of generating sufficient power at such a high frequency. Kühnold's transmitter put out one-tenth of a watt – not enough power to light a torch-bulb, let alone detect a distant object. That same autumn, however, the Dutch Philips company produced a valve able to generate 70 watts on a frequency of 600 megacycles – in itself a notable achievement at that time. The German rebuilt his device to use the new valves, and in January 1934 the Gema company began constructing his experimental radar.

On 20th March, Rudolph Kühnold set up his improved radar set on a balcony overlooking Kiel harbour. Dished reflectors behind the aerials were arranged to beam the transmitter's radiations at the battleship *Hessen* lying at anchor six hundred yards away. When he switched on his set, he observed no echoes at all from the *Hessen* as his sensitive receiver was swamped by signals coming from the transmitter, set up only six feet away. He switched the set off and moved the receiver further away from the transmitter. When he switched the device on again, he was able to make out quite clearly the echo signals returning from the battleship. Kühnold had rediscovered Christian Hülsmeyer's 'Telemobiloscope'.

The prototype radar was further improved, and set up at

the German Navy's research establishment at Pelzerhaken near Lübeck. In October Kühnold ran a demonstration for naval officials, and showed them echoes from a ship seven miles away. A small seaplane chanced to fly in front of the radar 700 yards away, and the scientist was able to point out that the aircraft's echoes were picked up by the set as well. The device greatly impressed the assembled officials, and the Gema company was awarded a development grant of 70,000 Reichsmarks, about £6,000.

Ten months later, Kühnold started using pulsed transmissions instead of continuous wave ones, an important improvement. By measuring the time lapse between pulses' transmission and return he was able to calculate the range of the targets. At this stage, his work was about four months behind that on radar being undertaken in Britain, and of course quite independent of it. In September 1935 Admiral Raeder, the German Navy's Commander-in-Chief, was shown the latest radar at Pelzerhaken, fitted to the 500-ton trials ship *Welle*. The latest set transmitted pulses on 600 megacycles and picked up echoes from coastlines twelve miles away and from other ships at five.

The Germans assigned their radar set the cover-name of '*DT-Gerät.*' Ostensibly, *DT* stood for *Dezimeter Telegraphie*, which would tend to link the system with the network of point-to-point wireless communication stations publicized by the German Post Office.[1] In the spring of 1936, the Gema company altered the frequency of *DT* to 150 megacycles, which extended its aircraft-detecting range to thirty miles. After further alterations, this set became the famous *Freya* radar operating on 125 megacycles, the most important German early-warning radar up to the middle of the Second World War. By the end of 1936, this radar had an aircraft-detecting range of fifty miles, and both the navy and the air force placed orders for it. When the German armed forces held joint manoeuvres in the Swinemünde area during the following autumn, an experimental *Freya* was one of the innovations, operated by Gema civilian tech-

1. For security reasons, the British had also given their radar a code-name, 'RDF', or 'radio direction finding'.

nicians. It stood on a small hill near Gölm, and caused a sensation amongst those in the know when it plotted aircraft flying sixty miles away.

The German Navy took delivery of its first *Freya* radar early in 1938. While this was a good aircraft early-warning set, the navy also required an accurate maritime radar set for ranging-in ships' guns. The Gema company had already produced such a set, code-named *Seetakt*, working on a frequency of 375 megacycles. The first prototype had gone to sea on board a trials ship in the autumn of 1937, and had shown a maximum range of nine miles on other ships with sufficient accuracy for gunnery. The pocket battleship *Graf Spee* carried an early model of *Seetakt* when she intervened in the Spanish Civil War in the summer of 1938.

By this time, the Gema company had a rival: the Telefunken organization had also entered the field of radar development, in 1936, and the *Würzburg* radar, produced at the end of 1938, was the impressive result: this was a small, highly mobile set with the ability to plot aircraft to within very fine limits at ranges up to twenty-five miles. The set worked on what was at that time an extremely high frequency – 560 megacycles – much higher than anything under development in Britain. It was this that gave it its excellent resolving power. *Würzburg* was in consequence the first radar device to come anywhere near to meeting the exacting requirements of the anti-aircraft gunners, who needed to be able to engage unseen targets. It was ordered into full production immediately. At the same time, the Telefunken company designed and built a small radar set for use on board aircraft, and this began trials in the summer of 1939 installed in a Junkers 52 transport aircraft.

How did all this compare with British radar development at the time of the outbreak of the Second World War? The *Freya*, the only German 'early-warning' radar, had a maximum range of 75 miles, gave a full 360-degree cover and was fully mobile; but it could not measure the actual altitude of approaching aircraft. Its nearest British equivalent, the Chain Home, had a maximum range of 120 miles and could determine an aircraft's altitude, but it could only gaze out over a fixed 120-degree arc, and its four 300-foot high transmitter aerials – each twice the height of Nelson's column – precluded mobility. It was not in the radar hardware that the British had established a lead, but rather in the way in which the information was used: only in the RAF Fighter Command did the means exist for channelling reliable and up-to-date information based on radar plots to fighter pilots by radio. The Germans made no attempt to perfect such a system prior to the war. Since they had then little to fear from hostile bombers, they logically concentrated their energies on more offensive developments – like the radio beams.

The German ground precision equipment, on the other hand, was now second to none: when war broke out, the Royal Navy had no equivalent of *Seetakt* for its ships – and would not have for two more years. The *Würzburg* was also considerably in advance of its nearest British equivalent, both in range and plotting accuracy. But the British had the lead in radar sets small enough to be installed in aircraft: they had two types, one for coastal patrol aircraft and one for night fighters, both on the point of entering service.

As Germany went to war in the autumn of 1939, she had eight *Freya* stations – two on Heligoland, two on Sylt, two on Wangerooge, one on Borkum and one on Norderey – covering the approaches to the short strip of German coastline between the frontiers of still-neutral Holland and Denmark.

The RAF was prohibited from attacking targets on the German mainland, where civilian lives and property would be endangered, so they sought out the German fleet in the Heligoland Bight. The first three daylight bombing attacks were inconclusive. But on the morning of 18th December 1939, twenty-four Wellingtons of Nos. 9, 37 and 149 Squadrons took off from bases in East Anglia and headed eastwards, led by Wing-Commander Kellett, commanding

officer of 149 Squadron. His orders were to patrol the
Schilling Roads, Wilhelmshaven and the Jade Roads, and
bomb any warships sighted at sea.

Just after midday, a *Freya* station on the pre-war holiday
island of Wangerooge picked up the Wellingtons approaching
at a range of seventy miles. The operator immediately
telephoned the fighter-aircraft station at Jever, but twenty
minutes elapsed before the first aircraft took off to engage
the bombers, by which time the Wellingtons were approaching Wilhelmshaven itself. It was a beautifully clear day.
The German fighter-pilots could see the diamond-shaped
British formation from several miles away, and sixteen Messerschmitt 110's and thirty-four Messerschmitt 109's went
into action. The British formation turned and headed for
home, but by then the German fighters were in hot pursuit.
Of the twenty-four bombers only ten returned to their airfields. The RAF had learned the hard way, as the German
Air Force was to learn during the Battle of Britain and the
Americans were to learn in 1943, that formations of bombers
unescorted by fighter aircraft could not survive in daylight
operations. In future, RAF Bomber Command would have
to avoid the fighter defences by going in under cover of
darkness.

The ban on air attacks on the German mainland was lifted
by Mr Churchill, newly appointed Prime Minister, following the German attack on Rotterdam on 14th May 1940.
By 4th June, RAF bombers had flown some 1,700 sorties
over Germany by night, losing only thirty-nine aircraft,
mostly in accidents. Compared with what Bomber Command was to achieve later in the war, these early operations were little more than gestures of defiance, but they
caused consternation in Germany. Had not Göring declared
that such a thing was impossible? The Reichsmarschall's
imagination had in fact been captivated by the *Würzburg*
radar, for all its many 'boxes with coils': here was a device
which would enable his guns to operate effectively under
all conditions, even the thickest cloud, which searchlights
could not penetrate. It was the success of the early *Würzburg* trials which had inspired him to make his much-quoted

declaration that the Ruhr would not be exposed to one single bomb dropped by enemy aircraft, after inspecting gun-defences in the Essen area in August 1939.

But the introduction of *Würzburg* gunlaying radar took longer than Göring anticipated, and without radar the gunners still had to seek out their targets using searchlights and an optical range predictor, and this was a difficult task. The standard anti-aircraft gun was the 88-millimetre Flak, which was capable of hurling an eighteen-pound shell to a maximum slant range of 9,000 yards; the nub of their problem was that the shell took twenty-five seconds to cover this distance, during which time its target aircraft would move nearly two miles. Radar was obviously going to be of the greatest importance for laying anti-aircraft guns, but *Würzburg* only began to enter service in the summer of 1940. Göring was not at all satisfied by this – while his reputation had suffered, the raiders had not.

There was no specialist night-fighter force in Germany to counter British bombing raids, so on 17th July 1940 Göring summoned Colonel Josef Kammhuber and ordered him to set one up. Kammhuber was 43 years old when he took on his new appointment. In his previous posts he had demonstrated that he was a methodical worker, displaying great drive tempered by sound judgment. He was to use these qualities to the full in his new job. By the end of July, his command comprised two flights (*Staffeln*) of Messerschmitt 110's and a few of the new Junkers 88 fighter-aircraft and a number of Messerschmitt 109's – about thirty-five aircraft in all. Backing this force on the ground was one searchlight regiment and a few *Freya* early-warning radar sets. Kammhuber was promoted to Major-General, and established his headquarters in a beautiful seventeenth-century castle at Zeist, near Utrecht. The night-fighter organization was subordinated to Colonel-General Hubert Weise, who was responsible for the air defence of the Reich.

At this time, night-fighting was in its infancy. Generally the fighters took off after receiving warnings of the approach of raiders from the early-warning radar stations on the coast, and the pilots orbited radio beacons until searchlights illuminated the bombers. Once they could see their targets, the fighters closed in for the kill. This system achieved successes, and was known as 'illuminated night fighting'.[2] There was one important drawback: all the searchlights were positioned near towns, so these tactics rarely achieved anything outside the target area. There were serious disadvantages to operating night fighters in the flak areas. The main problem was that of identification: if there was a mistake the gunners opened fire on the fighters. In war such risks sometimes have to be accepted if there are compelling reasons. But in this case there was no such necessity. Quite apart from the wastage in men killed and wounded, in aircraft destroyed and damaged, the system was unsatisfactory because the gunners spent their time shooting at fighters and the fighters spent their time evading shells, when both should have been engaging the bombers.[3]

General Kammhuber was quick to appreciate this, and made sweeping changes: he deployed his searchlights clear of the German cities – and therefore clear of the guns – in a belt which ran parallel to the coast, from Schleswig-Holstein to Liège. The raiding aircraft would all have to pass through this line to reach their targets. To ease the problem of identification still further, the belt was declared a prohibited area for all German aircraft not engaged on night interception, during the hours of darkness. Kammhuber's night-fighters, each patrolling a 'box' of the night sky in this line, could now concentrate their efforts on all unidentified aircraft crossing the belt.

Having worked out his tactics, Kammhuber set about equipping his force with the tools to do the job. Any system dependent on searchlights was a slave to the weather. What was needed was a night-interception technique not tied to the use of searchlights, but this in turn called for a radar set with which a ground controller could bring a night-fighter right up to its target, the enemy bomber. The *Freya* set was unsuitable, because it could not differentiate between single

2. *Helle Nachtjagd.*

3. The positive identification of aircraft in a confused tactical situation is still a major air defence problem.

aircraft flying close together; the 'blips' from the fighter and
the bomber merged on the screen long before the fighter
pilot was within visual range of his quarry. The *Würzburg*,
with its high frequency and better resolving power, could
do the job, and a number were ordered for the new night-
fighter force.

By the end of 1940, *Würzburg* production was at last
getting into its stride. For the gunners too it seemed the
answer to many problems. A radar-controlled 'Master'
searchlight was introduced: this searchlight, which seemed
to the bomber crews to have a distinctly bluish tinge, could
tilt straight over on to an enemy aircraft without first groping round for it in the darkness. One Wellington wireless
operator who flew thirty operations over Germany between
January and September 1941 later described: 'The Master
searchlights gave off a pale blue light and would point up
vertically as we approached. Then they would suddenly
swing on to a bomber and hold it, whatever evasive
manoeuvre was tried. Once illuminated by the Master, the
remaining searchlights in the group – generally four –
were switched on. They would cone the aircraft in blinding
white light. Then the Master would resume the vertical.
Once one was held in a cone, all hell broke loose. The only
way to escape was to dive, to pick up speed as quickly as
possible, and get clear of the area.' No RAF Bomber Command pilot will ever forget the chilling spectacle of a British
aircraft trapped in a searchlight cone, desperately weaving
in an attempt to escape before the guns found it.

In August 1941, Josef Kammhuber was appointed
General of Night-Fighters. He was at last beginning to secure the *Würzburg* sets he needed to equip the defensive
line to the west of the Reich: the line was now shaped like
a giant inverted sickle. The 'handle' ran through the middle
of Denmark from north to south: the 'blade' curved through
northern Germany, Holland, northern Belgium and eastern
France, to the Swiss frontier. The 'Kammhuber line' was
divided into a series of boxes about twenty miles wide. Each
box was equipped with one *Freya* and two *Würzburg* sets;
the former directed the latter on to aircraft. It was just

possible for *Würzburg* operators to seek out their targets unaided, but – because it had such a narrow beam – this took time. With bombers crossing the line at more than three miles per minute, time was of the essence. As soon as a raider entered the range of *Würzburg*, one set tracked it while the other followed the movements of the night-fighter patrolling that particular 'box'. The box's controller radioed 'intercept vectors' to the fighter pilot, and brought the two within visual range.[4]

Kammhuber positioned his radar stations immediately in front of the old searchlight belt. Thus the fighter pilots could try a radar-controlled interception first, and if this failed they could resort to the tried and tested illuminated night-fighting technique, with the advantage that the fighter was actually in pursuit of the bomber before the latter entered the searchlight zone. With radar to increase the effective 'depth' of the searchlight belt, Kammhuber was able to redeploy nearly half his searchlights to increase the length of the line. The reorganization of the defences of the Reich along these lines took a major part of 1941; by the autumn of that year it was complete.

Even so, the Kammhuber line had its deficiencies; for maximum efficiency, the *Würzburg* sets fell short of what was required. The trouble was that its range was so short that it was often impossible to find the attacking bomber's altitude in time for the night-fighter to reach it; and unless the interception was quickly completed, the enemy bomber passed out of *Würzburg* range unscathed. Ground reflections also made it difficult to follow aircraft flying below 6,000 feet. The Telefunken company's engineers set about overcoming these problems and during the spring of 1941 the German Chief of Air Staff, General Jeschonnek, was advised by the Director-General of Equipment that a new device was going into production while still hardly off the drawing-board, designed to meet these objections: Telefunken's engineers had in fact increased the *Würzburg* reflector-dish from ten feet to twenty-five feet in size; this had the effect

4. The Germans code-named this system *'Himmelbett'* – four-poster bed.

of narrowing the beam width, while more than doubling the range, enabling it to detect aircraft over forty miles away.

The new Telefunken radar was called the Giant *Würzburg*. Apart from its larger reflector and the static mounting, it differed little from the smaller version. It was ready for service in the winter of 1941, when it began to replace the standard *Würzburg* sets previously used for night-fighter

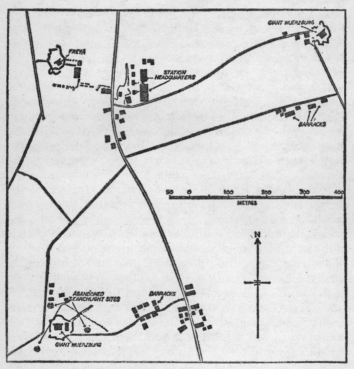

A *Himmelbett* Night Fighter Radar
Control Station

This is a typical station; it is in fact the one at Nieu-wekerken, Belgium, and was to become the subject of close scrutiny by British Intelligence.

control. The latter sets had of course started life as radar for gun-laying, and its form of presentation was quite unsuitable for directing night-fighters; the fault was perpetuated in the Giant *Würzburg*.

To convert the range-and-bearing information derived from the radar into a form in which it could be used by the fighter controller, the German air force developed the 'Seeburg table': located in the headquarters building of each ground-control station, this looked rather like a dais with two flights of steps leading up to a table at the centre. The table top consisted of a frosted glass screen, with a map of the area and a grid painted on it. Beneath this screen was a second table, round which sat two men operating light-projectors, one to project a spot of red light indicating the bomber's position, and the other to project a blue spot indicating the fighter's position, on to the screen above. Each man was connected to one *Würzburg* set by telephone. As the two coloured spots of light jerked across the frosted glass screen, a man at the top of the 'dais' followed them with a coloured wax crayon. The fighter controller could see the progress of the interception and broadcast instructions to the fighter aircraft by radio telephone. These 'Seeburg tables' became standard equipment in each box's ground control station.

Early in 1942, Hitler decided that with this influx of new electronic equipment General Kammhuber no longer needed the searchlights in his defensive line. He therefore ordered their redeployment. At the time, Kammhuber strongly contested this decision, but he later admitted that it was for the best. It forced the night-fighter crews to trust the ground controllers to guide them to their targets and when the men became accustomed to it the system was more effective than 'illuminated night-fighting' could ever have been. As we shall see, the moving of the searchlights was to cause British Intelligence a lot of trouble.

During the spring of 1942, the Kammhuber line was strengthened by the introduction of three new radar devices: the formula of using a larger aerial reflector to increase

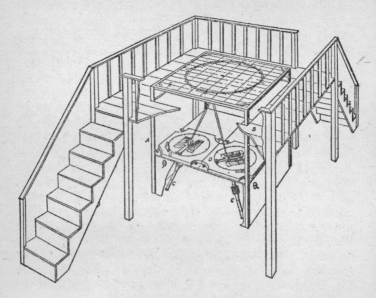

The 'Seeburg' Plotting Table

At the lower table sat two projector operators; one
was connected by telephone to the Giant *Würzburg*
radar tracking the fighter, the other to the radar
tracking the bomber. The projectionists each shone a
light – a red one for the bomber, a blue one for the
fighter – on to the corresponding positions on a grid
marked on the base of the frosted-glass screen above
their heads. At the top of the translucent screen the
paths followed by the lights were crayoned in, thus
marking out a representation of the tracks flown by
the two aircraft. Watching this process at the top of
the 'dais' was the fighter control officer; he passed
instructions to the night fighter pilot by radio, to bring
about an interception.

range, which had worked so well in the case of *Würzburg*, also succeeded with *Freya*. The result was *Mammut* ('Mammoth') built by the IG Farben company. It was basically an enlarged *Freya* with a reflector ninety feet wide and thirty-five feet high – about the size of a tennis court. The structure did not rotate, but found the direction of the target by 'swinging' the beam electronically through a limited arc of 100 degrees. The huge reflector squashed the beam into a narrow pencil, which could reach out to aircraft up to 200 miles away.

Like its smaller predecessor, *Mammut* could not measure altitude. The second new radar, *Wassermann,* was built by the Gema company, and this could give accurate information on the height, range and bearing of aircraft up to 150 miles away. It employed an aerial mounted on a rotating tower with a reflector 130 feet high and twenty feet wide. *Wassermann* was the finest early-warning radar to be produced by either side during the Second World War. The Germans installed *Mammut* and *Wassermann* sets along the west coast of occupied Europe, to give warning of the approach of raiders.

These two were basically early-warning systems, but the third of the new radars was of an entirely different character: *Lichtenstein* was an airborne radar, designed to be fitted to night-fighters to enable their crews to engage bomber targets even on the darkest night. Built by Telefunken, the *Lichtenstein* worked on 490 megacycles and had a maximum range of two miles and a minimum range of 200 yards or so: the minimum range at which targets can still be 'seen' is obviously an important parameter in night-fighter radar. That there is a minimum at all becomes evident from technical considerations: the radar transmits a pulse and while it does so the extremely sensitive receiver has to be 'switched off' otherwise it would suffer severe damage. The receiver cannot pick up echoes from targets nearby, because the transmitter is still radiating its brief pulse and the receiver is still 'switched off'. This dead distance between the radar set and its nearest 'visible' target is proportional to the length of the transmitted pulse. In fact the *Lichtenstein*'s minimum range of about 200 yards was very good for such an early radar.

General Kammhuber persuaded Hitler to give the airborne radar the highest priority in production and the first four night-fighters fitted with *Lichtenstein* arrived at the operational airfield at Leeuwarden in Holland in February 1942. Here, the device's shortcomings became clear: the entanglement of aerials and reflectors appended to the aircraft's nose acted as an air brake, impairing the handling characteristics and reducing the Junkers 88's top speed by five knots. Few pilots were willing to accept this handicap in exchange for the privilege of carrying a radar set of unproven reliability. Paradoxically, the main reason for this conservative attitude towards the new airborne radar was not its design but the high quality of the radar-assisted ground control system, which generally resulted in fighter aircraft being led right up to their targets. Some of the foremost aces refused to have anything to do with the new *Lichtenstein* and others followed their example. It became commonplace for radar-equipped night-fighter aircraft to stand idle, while those without flew on operations.

Only one officer, Captain Ludwig Becker, and his crew persevered with *Lichtenstein*. They found that when it could be coaxed into working – it still had its share of electronic teething troubles – it offered considerable advantages, particularly on very dark nights. As Becker's score of kills steadily mounted, the device began to gain acceptance elsewhere.

By the close of March 1942, when the RAF Bomber Command area offensive opened with fire-raising attacks on Lübeck and Rostock, the German defences were destroying on average four out of every hundred bombers attacking Germany by night. Of these 'kills', the fighter defences were responsible for two-thirds, and the guns for the remainder. By now there were four wings (*Geschwader*) of night-fighters, with 265 aircraft of which 140 might be available on any one night. The force's expansion was continuing: during April it accepted thirty-three Messerschmitt 110's, twenty Junkers 88's and thirty Dornier 217's – more than enough to compensate for any wastage. The Telefunken company had manufactured 275 *Lichtenstein* airborne radar sets already, and production had reached the rate of sixty sets a month.

On the ground radar side as well, there was a steady strengthening of the line: Kammhuber needed 185 Giant *Würzburg* sets to equip the whole of his defensive line. By the end of March 1942, Telefunken had delivered about half of these, and the remainder were following at the rate of thirty every month. The German defences were taking a steadily mounting toll of the British night raiders: but they had their Achilles' heel. Kammhuber's system depended for its success entirely on the *Lichtenstein*, *Freya* and Giant *Würzburg* radar stations, and upon the communications between fighter pilot and ground controller. All of these were vulnerable to interference in some degree. But first, British Intelligence had to penetrate the veil of secrecy which surrounded the German radar defences.

3

Discovery

The relief offensive for which Britain's badly harassed Allies have been begging for such a long time has confined itself to the landing of a few parachutists on the coast of northern France. The parachutists were soon forced to make a glorious retreat across the ocean, without having achieved any useful purpose.
– German News Broadcast, 28th February 1942

No man can guarantee success in war, but only deserve it.
– WINSTON CHURCHILL

It was now clear to RAF Bomber Command that a steadily increasing proportion of their aircraft were failing to return from operations over Germany. This presented the Command with a difficult problem, for only when it was known how these aircraft were being sought out and shot down could appropriate countermeasures be taken. Most of the Intelligence sources that had proven so useful during the 'battle of the beams' – crashed enemy aircraft, captured aircrew and the analysis of the beams themselves – were now denied them: the Bomber Command aircraft were being destroyed over enemy territory. In consequence, the exposure of the system created by General Kammhuber was to take more than two years of hard work, in which the highlight was a combined operation hazarding the lives of more than a hundred British soldiers, in February 1942.

In the absence of any firm evidence before the war, British scientists had regarded with scepticism the possibility that the Germans might also be working on radar. There seemed little doubt that the Germans were capable of building such equipment once they had the basic idea, since they were known to be advanced in radio high-frequency technique. But against this it was argued that the absence of any reports from Germany of 'high towers' – like those employed by the early radar sets in Britain – indicated that the Germans had done no such work; nobody had picked up any radar signals emanating from outside Britain, but this was probably because the radio monitoring service had concerned itself only with simple communications transmissions. Those organizations which did use the specialized radar receivers had neither the time nor the desire to hunt round the radio 'spectrum' searching for other people's radar transmissions.

The first clue that the Germans had been working on radar for some time was a brief mention in the November 1939 Oslo Report: the anonymous informer wrote of an aircraft detection system with a range of 75 miles, employing a chain of radio transmitters along the German coast. He had said that he did not know the operating frequency, but had suggested that it could easily be determined by the British listening service. He had also described a second type of aircraft detection device working on a frequency of about 600 megacycles and using a parabolic (i.e. bowl-shaped) reflector. During the first two British air attacks on Wilhelmshaven in September 1939, the Oslo Report had added, these radar stations covering the north-west coast of Germany had detected the RAF bombers at a range of 75 miles.

This last statement was greeted with some disbelief. The German fighters' reaction to these attacks seemed to have been very slow if such a long advance warning had been given; certainly the British fighter organization would have moved much faster.

During the six months that followed the Oslo Report, the British Scientific Intelligence Service was able to pick up very little information which could be connected with German radar. The main reason was that although it was technically efficient, the equipment was only in very limited use. The German strategy at the time was primarily offensive, and they concentrated on systems like their navigational beams for bombers; in consequence they tended to neglect radar, which was at that time an entirely defensive device. The British, who were on the defensive, were compelled to adopt the opposite course.

It was not until May 1940, with the RAF's night offensive against Germany just starting, that a prisoner mentioned that the German Navy had been experimenting with a 'radio echo' device for measuring the range and bearing of distant objects. The German anti-aircraft defences were said to be working on the same developments as well, but they were not so far advanced. He had then asked whether the RAF had an aircraft detection system similar to that used by the German air force: from the description, the German system bore unmistakable similarities to the British coastal radar chain, and it had also been set up shortly before the war under conditions of great secrecy. There was much here that agreed with the statements made in the Oslo Report.

On 5th July, shortly after the reception of the first *Knicke-bein* beam signals by a British aircraft, one of Dr R. V. Jones' Intelligence sources passed to him the gist of a secret German air force report dated a week earlier: it stated that German fighter aircraft had been able to intercept British reconnaissance planes on that day because of information from the '*Freya-Meldung*' – the *Freya* warning. This seemed to confirm that the Germans had some form of air-craft detecting system: perhaps some clue as to the principle on which *Freya* operated could be found in the RAF's re-cords for that day: but the aircraft had observed nothing unusual, so Jones drew a blank. Of the name *Freya*, how-ever, there could be no doubt. On learning of Jones' in-terest, the same source mentioned that a *Freya* site was in operation at Lannion protected by a battery of light anti-aircraft guns. Lannion was a small village on the northern coast of Brittany. This report underlined the importance of the *Freya* system, for the Germans had entered the area only three weeks earlier.

There were two obvious ways in which to follow this

report up: the site should be photographed from the air as soon as possible, and a listening watch should be maintained for any signals that could be traced to it. In fact Jones adopted a further approach as well, typical of the methods adopted by the strange craft of military Intelligence: he did some research into the mythological background of the German code-name, *Freya*. Those who had named the device could hardly have chosen a more fruitful goddess, but few of her attributes seemed to bear any possible relation to the problem in hand: Freya was the Nordic goddess of Beauty, Love and Fertility. Dr Jones researched even deeper. He learned that Freya's most prized possession was an exquisite necklace called Brisingamen; to acquire this, she had had to sacrifice her honour and be unfaithful to the husband she loved. All this seemed to ring faint bells. Heimdal, the watchman of the gods, guarded Brisingamen for her, and Heimdal could see one hundred miles in every direction, by day or night. So that was it! Dr R. V. Jones cautiously reported to the Chiefs of Staff:

It is unwise to lay too much stress on this evidence, but these are the only facts that seem to have any relation to our previous knowledge. Actually Heimdal himself would have seemed the best choice for a code-name for RDF [radar] but perhaps it would have been too obvious ... It is difficult to escape the conclusion therefore that the *Freya-Gerät* is a form of portable RDF.

Freya may possibly be associated with *Wotan* – she was at one time his mistress – although it would have been expected that the Führer would have in this case chosen Frigga, Wotan's lawful wife.

All this was good stuff, but it awakened nagging suspicions in the minds of the War Cabinet. Had the Germans possibly captured intact one of the radar sets left by the British Expeditionary Force in France during the Dunkirk evacuation? How else could the Germans now have an operational radar system. On 7th July, Mr Churchill minuted General Ismay, his Chief of Staff, about this matter, and he pinned one of the feared red tags with bold black lettering – ACTION THIS DAY – to the minute:

Ask the Air Ministry whether any RDF stations fell intact into the hands of the enemy in France. I understand there were two or three. Can I be assured that they were effectively destroyed before the evacuation?

General Ismay made inquiries at once, and he replied on the same day that one radar transmitter had indeed been left behind by the RAF at Boulogne; but this had been most thoroughly destroyed beforehand, and it was very doubtful that the Germans had managed to extract any information from it. One of the army's gun-laying radar sets might have been captured in a damaged state, but all the others were carefully destroyed; again a complete gun-laying radar had been in the *Crested Eagle* when she was run ashore at Dunkirk, but a naval party had been put on board to destroy it: 'It is not quite certain whether this destruction was effective.' Ismay mentioned that there was a possibility that radar information had reached the Germans from the French, as the latter had had complete information about the British developments and there was no news as to what they had done with it: 'In the present circumstances it is extremely difficult to find out.'

From German sources, we know that the Germans did in fact capture an intact British radar set near Boulogne. Far from being impressed with their find, they regarded it as an extremely crude piece of work, much inferior to their own set, *Freya.*[1]

On 14th July, Jones received an Intelligence report of a second *Freya* station in operation, this time at Cap de la Hague, the north-western tip of the Cherbourg peninsula. Nine days later this station played an important part in the operation during which German dive-bombers sank the destroyer HMS *Delight*. At the time, *Delight* was about

1. At this time the mobile British sets – it was one of these that the Germans captured – were greatly inferior to their static counterparts.

twenty miles south of Portland Bill. As she had never been closer than sixty miles from the enemy radar station, and had had neither aerial support nor balloons to reveal her position, the action provided important confirmation for the reports Jones had received: the *Freya* set obviously gave at least as good a cover on low-level targets as did the very latest British radar sets. Finally, in the second week in August, as the Battle of Britain was about to begin, Dr Jones received the text of a secret German report on *Freya* which made it quite clear that the device was designed to work in conjunction with the German fighter defences.

The attempts to locate *Freya* on aerial photographs of the two sites now reported, at Cap de la Hague and at Lannion, met with no success. The photographs were taken from aircraft flying at 30,000 feet and gave a definition only just sufficient to enable 300 foot *Knickebein* turntables to be picked out. All that could be said for certain was that *Freya* must be smaller than that.

In the meantime, a radar expert from the Telecommunications Research Establishment, Mr Derek Garrard, began the search for actual radar signals. He filled his car with borrowed receiving equipment and drove down to Dover to look for unusual transmissions. His initiative was rewarded with immediate success, although it was not connected with *Freya*: he picked up radar transmissions on a frequency of 375 megacycles, and he was able to link these with the shelling of British convoys passing through the Straits of Dover by German coastal batteries situated near Calais. The radar signals came in fact from a *Seetakt* set – like the one whose aerials a British naval Intelligence officer had examined on the *Graf Spee* many months before. This find caused some commotion among radar experts in Britain: of those prepared to admit the possibility of the Germans having radar, very few would accept that it might be actually superior to that in Britain. Yet here was a German set – working on a frequency so high as to be only just usable in Britain – that was actually in service directing coastal gun batteries. If the Germans had learned their radar techniques from a British set captured after Dunkirk, they had indeed applied their newfound knowledge with the most re-markable, dispatch.

In the autumn of 1940 improved reconnaissance cameras became available, and these brought about a great improvement in the quality of British aerial photography. The effect on the hunt for the radar stations was immediate: on Sortie No. *H/458* on 22nd November, a high-flying Spitfire returned to Britain with photographs of unprecedented clarity, showing the little village of Audeville at Cap de la Hague: just to the west of the village were two unexplained circles side by side, and each measuring some twenty feet across, together they looked like a pair of opera glasses laid lenses down. Dr Charles Frank, a physicist who had just joined Jones' staff, examined the photographs through a stereo-scope, and he noticed that two consecutive frames did not form a perfect stereoscopic pair as they should have done: a shadow associated with one of the 'circles' had changed slightly in the nine seconds between the two exposures. During that time, a thin wide object on top of one of the circles had rotated through ninety degrees. In each case the shadow was about two millimetres long, but whereas in the first picture it was about 0·1 millimetres wide, the width in the second was two millimetres. The difference was hardly greater than the resolving power of the photographs, but it was enough to establish a vital step in the hunt for *Freya*. The next move would be to lay on a low-level reconnaissance sortie to take a closer look at the 'opera glasses' and their superstructure.

Unfortunately the matter had to rest for several months, as Britain, still threatened by invasion, had more urgent demands to make of her photographic reconnaissance Spitfires.

In the interval, an Intelligence report reached Britain about a second German aircraft detection device, called '*Würzburg*'. It appeared that one *Freya* and one – perhaps two – *Würzburg* sets were to go to Rumania, and two *Würzburg* sets were to be sent to Bulgaria; all had been allocated for coastal defence units. Perhaps, Jones reasoned, this represented the minimum necessary to give continuous cover along these two nations' coastline; if that was so, it was easy to calculate that the *Freya*'s maximum range must be at least 57 miles, and that of *Würzburg* must be at least 23 miles. Jones did not yet know of course that the two radar systems were complementary to one another. But though his initial premise was thus wrong, the maximum ranges he calculated for the two systems were remarkably accurate. As a result of his efforts so far, Jones was now able to resolve two distinct types of German aircraft-detection apparatus, although neither had yet been clearly seen, or heard at all.

It was not until 16th February 1941 that a reconnaissance Spitfire was able to go in at low level to look at the 'opera glasses' at Audeville. This time, the Spitfire proved too fast for the automatic camera – the circular objects were missed in the gap between two successive frames. The pilot did manage to catch a glimpse of them as he swept past. He thought that they looked like some sort of gun position. A second low-level sortie flown by Flying-Officer W. K. Manifould six days later was completely successful. He returned with a magnificent close-up oblique photograph of the objective: each of the circles was clearly shown to be surmounted by a rotatable aerial-array. This was unquestionably a radar station.

Even as Manifould's photographs were being processed and readied for detailed examination, the second orthodox Intelligence approach was also bearing fruit in dramatic fashion: a listening station in southern England recognized pulsed signals transmitted from the site at Audeville – a *Freya* station – on the very high frequency of 120 megacycles. These had been heard for some time, but hitherto it had been assumed that they emanated from the RAF fighter aircraft using new VHF radio sets. Only when Derek Garrard – by now a member of Jones' staff – examined the trace of the signals on a cathode-ray tube was their full significance recognized: they were radar transmissions coming from the other side of the Channel. Garrard plotted the bearings on their source, and found that not only did they originate from the Audeville site at Cap de la Hague, but they came from the Dieppe and Calais areas as well.

Within four hours of each other, Dr R. V. Jones received the news of the low-level photographs and the radio interception of *Freya* signals, after a hunt which had lasted over a year.

In spite of the mass of accumulated evidence, the fact that the Germans had developed radar equipment had still not been completely accepted by those in authority. Air Marshal Joubert had convened a meeting on 23rd February 1941 to discuss the fundamental question of whether the enemy had radar at all; Dr R. V. Jones attended, taking with him the remarkable Audeville photograph of the day before. This and the news of the interception of the site's signal settled the argument there and then. At first Joubert suspected that Jones had kept the photograph to himself, perhaps in order to manoeuvre him, Joubert, into a false position. It was only after Jones had drawn his attention to the date on the photograph that Joubert accepted that this was not so.

Within a month of the finding of *Freya*, new radar signals were picked up, this time on a frequency of 570 megacycles. A special radio-reconnaissance unit, No. 109 Squadron, had begun to operate during the spring, flying sorties in an attempt to find German radar transmissions. On 8th May one of this unit's 'ferret' Wellingtons was sent on a circuitous route round the Cherbourg peninsula and Brittany to seek out the sources of the new radar signals, and its crew was able to obtain rough fixes on nine sites in all.

That was not all, however. They noted 'searchlight activity consisted in the sudden switching on of a powerful light directed towards the plane with good accuracy. This seemed to be used as a pointer for other lights and may have been controlled directly by the RDF instrument.' This was the start of what was to become a great controversy.

In the meantime, information on the *Freya* radio stations was coming in thick and fast. Bearings were taken on all signals heard, and before the end of October 1941 it was possible to plot the positions of no fewer than twenty-seven of these radars – strung out along the coast between Bodo and Bordeaux. No. 109 Squadron's aircraft also brought back scores of fixes on 570-megacycle transmitters, but the device was evidently very small for it defied all attempts to photograph it.

One of the most spectacular pieces of Intelligence to be obtained at this time was a strip of cinematographic film showing a *Freya* station actually in operation, and German crews tracking aircraft targets.

Less spectacular but far more important from the Intelligence point of view was the discovery that the wirelessed plots from the *Freya* units could be overheard in England. As they tracked aircraft, each of the German radar stations radioed distance-and-bearing reports to a central air-defence headquarters. The 'code' used to pass this information was a relatively simple one, and it was easily broken. For example, a listening station in England picked up a string of Morse characters on 10th October 1941, reading:

$$MXF = 114011 = 14E = X = 254 = 36 = +$$

It was deduced that MXF was the radio call-sign of that particular *Freya* station; 11 40 11 was the time in hours, minutes and seconds; $14E$ was the serial number of the transmission and X was an indication of the number of aircraft in the plot (X represented one, Y several and Z many); 254 was the bearing of the plot measured in degrees from the radar station and 36 was the range in kilometres.

This German leak was tapped to the full by British Intelligence: reconnaissance aircraft, maintaining an accurate record of their flight path by photographing the ground over which they flew, were sent out for the *Freya* stations to track. Listeners in England then tuned in to the German reports on the aircraft's progress; the decoded positions of the radar plots were obtained, and the bearings given back-plotted on a map. A number of radar stations was pinpointed in this way.

While all this was going on, No. 109 Squadron was bringing back scores of fixes on 570-megacycle transmissions,
but the device still defied all attempts to photograph it. It
must clearly be smaller than *Freya*. From agents' reports
Jones learned that these transmissions were connected with
a fire-control device known as the *FMG*. Perhaps *FMG*
stood for *Flakmess-Gerät* – a 'gun-range-measuring device?'[2] Towards the end of 1941, news arrived that four
FMG sets were operating in the Vienna area, of all places:
unless Vienna – a city of great beauty but comparatively
little military importance – was a radar equipment depot
it seemed reasonable to infer that the *FMG* existed in considerable numbers.

In the meantime Jones had received through the United
States – at that time still a neutral – a photograph taken from
the United States embassy in Berlin. The embassy over-
looked the Berlin Zoo district, and here the Germans had
built a huge concrete flak tower: clearly visible on top of
it was a radar aerial, of a type unknown to Jones. The
photograph showed a dish-shaped openwork structure, obviously a long way away, poking out above some trees.
Unfortunately there was nothing nearby which could be used
to give scale to the object. A few weeks later, a Chinese
scientist reported that he too had seen the device, which he
described as paraboloid more than twenty feet in diameter,
which could be both rotated and elevated; he thought it
must be used to direct the anti-aircraft guns with which it
shared the tower. It was clear to Jones that this could not be
the 570-megacycle set which the radio-reconnaissance unit
had shown to exist in such profusion in German-occupied
Europe; had that been so, the large aerial structure would
have been spotted on aerial photographs long before this.

In fact the Berlin Zoo photograph showed one of the
first Giant *Würzburg* radars to enter service. Jones still had
no pictures of the small *Würzburg*, a set which now existed
in far greater numbers, but the hunt was already nearing

2. The *FMG* actually stood for *Funkmess Gerät* – 'radio measuring device', a German term for radar. Jones was looking for a code-
name where none existed.

its end. Late in November 1941, Charles Frank was examining a medium-level photograph of the *Freya* station at Saint-Bruneval on the northern coast of France near Le Havre, when he noticed that a track had been trodden out along the cliff-edge: it ran from the *Freya* aerials towards a large house which seemed to be some sort of headquarters building. Just before it reached the building, the track swung round to the right, and ended at a small black object about half-way between the house and the cliff, with its sheer drop to the sea below.

Somebody had considered it worth while to tread out a path from the main radar station to this object. Could it be that the black object played some part in the working of the *Freya*? On 3rd December, Flight-Lieutenant Tony Hill, a reconnaissance pilot, chanced to visit the photographic interpretation centre at Medmenham in Buckinghamshire. Hill had come to discuss the low-level photography of the German radar sites with Squadron-Leader C. W. F. Wavell, and because Wavell knew of Frank's special interest in the Bruneval radar site, he mentioned the mysterious black object to Hill. On the following day, Hill took off in his Spitfire to go and have a look at Bruneval on his own initiative. He swept in low over the cliffs, and was past the emplacement and out over the trees behind it before the startled defenders knew what had happened.

On his return, Hill discovered that his camera had failed to function properly.

It was a cruel blow. But he had seen the device clearly himself – it looked like 'an electric bowl-fire and about ten feet across'. If Hill was right this was almost certainly the elusive source of the 570-megacycle transmissions. On the following day, Hill bravely repeated his performance. The photographs he brought back to England were among the classics of the war: they showed the radar device exactly as he had described it, like an electric bowl-fire about ten feet across. This seemed to be the device which had been mentioned in the November 1939 Oslo Report: certainly the parabolic reflector was there. But it still remained to be proved that the radar transmitted on 570 megacycles, and until this had been proven countermeasures could not begin.

It is difficult to establish who first suggested that it might be a good idea to pilfer the 'bowl-fire' from Bruneval. The notion was such an obvious one – the radar station was within two hundred yards of the coast – that the idea may well have occurred to several people at once. At all events, such an operation was receiving detailed consideration by the beginning of January 1942. It soon became clear that a commando raid from the sea would be doomed to failure: the Bruneval site was situated at the top of high cliffs, and protected by a sizable German garrison. Even if the raiders could have fought their way to the top of the cliffs without suffering crippling casualties, it was unlikely in the extreme that they could do so before the defenders destroyed the radar set.

The naval Commodore in charge of Combined Operations, Lord Louis Mountbatten, suggested that parachute troops should be used instead. On 21st January the Chiefs of Staff agreed to this, and ordered that one company of paratroops, sufficient light naval craft to evacuate the force by sea and an operational squadron of Whitley bombers to transport the paratroops be made available. The venture was given the code-name BITING.

'C' company of the Second Parachute Battalion was chosen for the operation. The unit began training for what they were told would be 'a special demonstration exercise, which will probably take place in the Isle of Wight, and the whole of the War Cabinet will be there to see it'.

While the military side of the operation was proceeding, Dr Jones devoted some thought to the points in which Intelligence would have a special interest during the attack. By this time he had learned that the 'bowl-fire' was in fact the radar set called *Würzburg*, and that such a set had been erected near the Bruneval *Freya* station – this latter confirmation had come in an agent's report. What still remained to be proven was that *Würzburg* was the source of the 570-megacycle transmissions: for this to be resolved it was necessary for somebody to remove the aerial from the centre of the 'bowl'. In case all else failed, a radio receiver was to be carried in one of the naval craft dispatched to evacuate the raiders from the beach, and this receiver would be used to establish whether any 570-megacycle transmissions could be heard coming from the station.

The next objective was the capture of the *Würzburg*'s receiver and its associated presentation, or 'display', equipment. These would reveal any anti-jamming circuitry the Germans had built into the set. The transmitter was also wanted, so that British scientists could form some impression of the German techniques on frequencies as high as 570 megacycles. Dr Jones also asked for two prisoners to be taken, radar operators if possible: then information could be obtained on the German methods of operating the radar and aircraft reporting. Finally, as all German signals equipment bore highly informative labels and inspection-stamps, and a great deal of useful general Intelligence could probably be obtained from these, Jones requested that should the various units prove impossible to unlodge, at least their labels should be torn off for him.

The actual dismembering of the *Würzburg* equipment would be undertaken by seven men of the Royal Engineers, commanded by Lieutenant D. Vernon. There was to be an eighth man in the team who was not a soldier at all: Flight-Sergeant C. W. H. Cox, an RAF radar mechanic, was to go along should any specialist knowledge be required. The dismantling party was given a British gun-laying radar – the nearest equivalent to the *Würzburg* – on which to practise, and they demonstrated considerable skill. Vernon and Cox were given a special briefing on the expected character and layout of the German set.

According to the raid's planned time-table, they would have barely half an hour with the apparatus: they had to be trained to make sketches and photographs of the equipment in this time, and then to dismantle it systematically, beginning at the aerial and working backwards through the receiver to the presentation gear.

By the fourth week in February all was ready, but now the weather intervened. On the evening of the 24th the weather was unsuitable, and it was again unsuitable on the 25th and the 26th. The timing of the raid was extremely critical: the attack had to take place on the night of a full moon, but at the same time the tide would have to be on the rise, for otherwise the assault craft would be left stranded on the beach. The 27th was the last possible date for a month or more. Fortunately the weather forecast was favourable for that evening.

Thus informed, the Commander-in-Chief Portsmouth, Admiral Sir William James, signalled:

Carry out operation BITING tonight 27th February.

Just before ten o'clock that evening, twelve converted Whitley bombers of No. 51 Squadron took off from Thruxton near Andover, loaded with 119 paratroops. On board each of the aircraft, the troops sat huddled together, cold and uncomfortable in their parachute harness. One of them wrote: 'The mugs of hot tea (well laced with rum) we had drunk before taking off began to scream to be let out. In that restricted space and encumbered as we were there was, alas, no way.'

It was a two-hour flight out to the dropping zone. At a quarter-past midnight on the morning of the 28th, the first sticks of paratroops leapt from their aircraft. Some ten seconds later the men tumbled down on the carpet of virgin snow, six hundred yards to the south of the radar site. They hastily threw off their parachute harness and cocked their weapons. Each man crouched low, ready to fight for his very life, but the half-expected rattle of enemy small-arms fire did not come. Their arrival had passed unnoticed. The sound of the Whitleys' engines faded into the night, leaving the men feeling very lonely and vulnerable.

Slowly they stood up and started to move. The men assembled in small groups and there were whispered orders. The next move was not a warlike one – for that tea just had to go. Major Frost, the force's commander, later wrote that this 'was certainly not good drill, as now was the time when a stick of parachutists was most vulnerable . . . but at least it was a gesture of defiance!'

The soldiers formed up into assault parties. One, under Frost himself and comprising fifty men including the dismantling party, moved towards the radar site and the house nearby. A second under Lieutenant Timothy took up covering positions, to screen the force from attacks from the landward side. The remainder made off to secure the beach and the escape route. Frost's men silently surrounded the *Würzburg* – its silhouette stood out sharply in the moonlight – and the nearby house. If the house was some kind of headquarters there might be strong resistance. Frost himself stole round to the front door with a small party. Satisfied that all was ready, he gave the signal for battle to begin – a long shrill blast on his whistle.

With four men at his heels, Frost burst into the house. It was all something of an anti-climax. The sole German present died trying to defend one of the upstairs rooms. Outside, a small battle was in progress, the almost continuous automatic fire being punctuated by the crashes of exploding hand-grenades. Within minutes all German resistance round the *Würzburg* had been overcome. But as suddenly as this fight had ceased, another had begun: there were about a hundred German troops in the area and Timothy and his covering force soon found themselves hotly engaged.

In the battle round the *Würzburg* five of its six operators had been killed. The sixth made off into the darkness, but in his haste he lost all sense of direction and stumbled over the cliff. Fortunately for himself he managed to grab hold of a projecting rock. With some difficulty he climbed the fifteen feet to the top, only to find himself being helped over the edge by one of the British paratroopers. He was in no position to offer any resistance, and quietly surrendered.

With the *Würzburg* secured, Lieutenant Vernon scrambled up and examined it with the aid of a hand torch. He began to photograph the aerial – an action no sooner made than regretted, since the blaze of light from the exploding flashbulbs attracted bullets from several directions. Vernon summoned the remainder of his team and ordered one sapper to saw off the aerial element while the remainder removed the radar components in the operating cabin. The aerial came away easily, but the main radar equipment defied several attempts to dismantle it with screw-drivers. This was no time for finesse – the bullets ricocheting from the cabin's walls were real enough. Crow-bars were brought into play, and the radar set gave up the unequal struggle. The units were ripped out of the console. The dismantlers had been at work for barely ten minutes of the planned thirty, when Major Frost saw three lorries approaching with headlights full on. Almost certainly these were German reinforcements. If the enemy brought into action any weapons heavier than the rifles and machine guns they were already using, his force would be at a severe disadvantage. Frost decided to settle for whatever the dismantling party had managed to secure and ordered the force to withdraw to the beach. But the actual shoreline, it was now learned, was still in German hands.

What had gone wrong? Of the forty men detailed to secure and hold the escape route, half had landed two and a half miles from their planned dropping-zone. The senior officer present, Lieutenant E. Charteris, took a quick bearing on the lighthouse at Cap d'Antifer and worked out his position. His men then set out for Bruneval at that Red-Indian-like lope – a cross between a fast walk and a run – which is the gait of the trained paratrooper. As they moved in towards the sound of gunfire, the soldiers saw enemy patrols going in the same direction. One of the German soldiers attached himself to Charteris' force by mistake, and the subsequent explanations resulted in his death. Charteris and his men arrived at the top of the beach just as Frost was organizing his own force to storm it. The combined assault teams rushed the German strongholds in their path.

Now it was a quarter past two. The beach was in British hands, and on it lay the wounded, the German prisoners and the pieces of equipment ripped from the *Würzburg* site. Frost had done almost everything that had been asked of him. He told his signallers to call in the naval assault craft to evacuate his raiding force. On the cliffs on either side, the presence of German forces was becoming increasingly evident.

After a few minutes, the signallers reported that they had no success in contacting the boats. As a last resort, Frost himself fired red distress flares. Then, as Frost wrote, 'I moved off the beach with my officers to rearrange our defences. It looked as though we were going to be left high and dry, and the thought was hard to bear.'

Just as his troops began to take up positions for a final stand, Frost heard a cry: 'Sir – the boats are coming in. The boats are here! God bless the ruddy Navy, sir!' Frost looked back and saw that six snub-nosed assault craft had slid to a stop on the beach. With a sigh of relief he ordered his men to embark. From the boats themselves, covering fire was poured on to the Germans at the top of the cliffs. First on board were the men of the dismantling party – carrying the precious pieces of the radar's equipment. Then came the prisoners and the wounded, and finally the rest of the parachutists followed. The landing craft backed away from the shore, while the brisk exchange of fire continued until long after the boats were clear.

Safe aboard an assault craft, Major Frost learned the reason for the delay. While he had actually been signalling, a German destroyer and two E-boats had passed within a mile of the small British flotilla. Fortunately they had noticed nothing. Frost also learned that the pieces of the radar set secured by his men were almost exactly what had been needed. Mr D. H. Priest, a Telecommunications Research Establishment engineer who had received a temporary commission as a Flight-Lieutenant for the occasion, had examined the booty on the boat. Had the coast been sufficiently clear, he would have landed and looked over the actual *Würzburg* site, but this was not possible now of course. By the time dawn broke, the force of returning assault craft was only fifteen miles off the French coast. Several Spitfires arrived overhead, and these escorted the raiders back to England and safety.

The Bruneval operation had been successful on almost every count: the men had captured most of the radar and three prisoners, one of whom was a radar operator. The paratroops' own casualties numbered fifteen – two dead, seven wounded and six missing. Frost and Charteris each received the Military Cross for their parts in the attack; two paratroop sergeants and Flight-Sergeant Cox were awarded the Military Medal. The dismantling party had done extremely well in the brief time available for pulling the *Würzburg* set apart: the units they had brought back included the receiver, the receiver amplifier, the modulator – the device controlling the timing within the radar – and the transmitter; in addition, they had the sawn-off aerial element.

The only unit which Jones had requested and not received was the presentation equipment. There had been no time to rip this away before Frost gave the order to withdraw.

If – as had nearly happened – the boxes had proven impossible to tear away from the mounting, Jones would have had to make do with the labels alone, so it is interesting to see what he would have been able to learn from them alone: the labels showed that the set had been constructed by Telefunken, a company with its main factories in the Berlin area. The works-numbers were particularly interesting: from previous experience with similar serial numbers, Jones deduced that that allocated to the first production model of each component was 40,000. The earliest number he found in the captured units was 40,144, and the latest was 41,093; this suggested that the total number of sets of components manufactured by the date of manufacture of the last item was 1,093. The earliest inspection date, early November 1940, was stamped on a part of the transmitter; the latest, 19th August 1941, was on the aerial. This did not necessarily mean that 1,093 complete *Würzburg* sets had been turned out by the latter date, since a substantial proportion of the component units would have been required as spares. It was a principle of German design that servicing should be facilitated by making complete units readily replaceable; the defective units were returned to a central repair depot.

Assuming that fifty per cent of radar production went into spares, Dr Jones reckoned that some five hundred sets were available by August 1941, at which time production was probably running at about one hundred per month. All in all, the raid was particularly satisfying for him, as the Intelligence gleaned from the venture served either to confirm or add to the previous picture, and no part of his picture had had to be modified or discarded. It was an important long-term effect of the raid that it gave British Intelligence added confidence in the accuracy of its information.

Scientists at the Telecommunications Research Establishment at Swanage made a thorough examination of the pieces brought back from Bruneval. They noted that the equipment was 'straightforward and in no respect is it brilliant . . . On the other hand it must be remembered that the equipment was made in 1940 and designed in 1939 or earlier.' That was a time when British radar had certainly not been sufficiently advanced to produce a set working on 570 megacycles with a range of 25 miles. The German radar did not carry any specific anti-jamming circuitry; but on the other hand it could be tuned over a wide range, a feature which would make electronic jamming very difficult.

During the subsequent interrogations, one unhappy Bruneval prisoner mentioned that he had been on leave in Germany the month before the raid. While at home, he had mentioned to his wife that his station was so isolated that the British might easily make a raid and capture it. Was his wife, he wondered, perhaps a British agent?

One German paratroop leader stated after the war that the Bruneval operation was outstandingly the best, both in conception and execution, of all the British commando raids. It certainly gave the British people a badly needed boost in morale. After the Bruneval horse had been 'rustled', the German local defence commander made sure that the door was well and truly shut: the remains of the *Würzburg* were removed and a new one was set up in the main *Freya* compound. Within a few weeks the latter had been effectively covered with a barbed wire entanglement which exceeded in complexity anything hitherto seen at a military site. Other radar stations followed this lead. Soon every German radar station near the coast had become conscious of its vulnerability, and had surrounded itself with barbed wire.

This greatly helped Jones and his staff: there were several sites which were suspected to contain *Würzburgs*, but in each case the existing photographs had failed to show them. The Germans had now obligingly ringed many of them with barbed wire – which showed up with great clarity on the aerial photographs – and thus confirmed the suspicions.

It was not only in Germany that the repercussions of the Bruneval operation were felt. The raid's success highlighted the golden opportunity that was offered to the Germans if they were to attack the Telecommunications Research Establishment, hub of Britain's work in the field of radar; the TRE was near Swanage, on the south coast of England. During the spring of 1942, it became very conscious of its own vulnerability and spurred on by rumours of an impending German paratroop attack the research centre was moved to a safer area.[3] Malvern College, unassailable except by full-scale invasion, became the new home of telecommunications research in Britain.

Two weeks before the Bruneval raid, an event of great significance to the development of the radio offensive against German radar had occurred. The German battle-cruisers *Scharnhorst* and *Gneisenau* ran the gauntlet of British defences covering the Straits of Dover, and successfully reached German ports. It was a great blow to the British reputation as a naval power and a board of inquiry was set up to find out how the German warships had been able to move in broad daylight less than twenty miles from the coast of England.

One answer was that the Germans had used jamming transmitters to 'blind' some of the radar stations keeping

3. The rumours had no substance. The Germans had not in fact located the research centre. After the war, the German air force's senior officers were unanimous that it had just not occurred to them that the British might be so foolish as to place such an important target anywhere but in a safe area.

watch on the Channel. While this was not the first time that the Germans had used jamming against radar – they had made half-hearted attempts to jam the radar system during the Battle of Britain – the new development was of great strategic importance. The British had been reluctant to prompt all-out radar jamming, as they had the most to lose in an exchange. But now the Germans had begun jamming in earnest, with no prompting from anybody. They had demonstrated that they had a radar-jamming organization tested and ready, and the British radar system was highly vulnerable to this jamming.

The British night-fighter control radars and the airborne-interception sets, the AA director radar and the radar directing the searchlights – all operated in the same part of the radio frequency spectrum, around 200 megacycles. A heavy German jamming effort here would pay handsome dividends should there be a resumption of the night Blitz. The escape of the German warships from Brest greatly assisted the British cause by highlighting this fundamental weakness. New radars working on a wider range of frequencies were rushed into service; never again would the Germans have the chance to knock out the radar system with so little effort.

At the same time this incident served to draw attention to the requirement for an active British jamming organization. Dr Cockburn and his team had been 'champing at the bit' to upset the German radar, and now they were giving a free hand to produce a whole range of jamming devices. For technical reasons involving its high frequency, it was difficult to use straightforward 'noise' jamming against the *Würzburg* radar; however, Cockburn's team devised a very effective weapon for use against the *Freya* radar, which worked on a much lower frequency. The new equipment received the pulses from the German *Freya*, which was essentially an early-warning radar, amplified them and sent them back; it was code-named 'Moonshine'. Because of these apparently huge 'radar echoes', the effect of this device was to produce on the *Freya*'s screen a display like that from some huge object or from several small ones in close proximity. By mounting 'Moonshine' in an aircraft it was possible to simulate the radar echo of a force of several aircraft fly-ing in formation. It was a useful method of drawing German fighters away from areas to be attacked.

A special unit, No. 515 Squadron, was formed in April 1942 to use 'Moonshine'. The squadron was equipped with nine Defiant aircraft; these two-seater planes, obsolete as fighters, were the only high performance aircraft readily available. The installation of the new 'spoofing' equipment began immediately, and by August it was complete. On the 6th, the device underwent its first operational trial: eight Defiants, orbiting south of Portland, 'Moonshined' the Ger-man early-warning radar screen. Some thirty German fighter aircraft – the entire air defences of the Cherbourg area – took off to meet the 'threat'. Eleven days later the 'spoofer' was used to divert the enemy's attention from a real attack. While the 'Moonshine' Defiants and a hundred supporting aircraft assembled menacingly over the Thames, American Flying Fortresses attacked Rouen, escorted by a large num-ber of fighter aircraft. The German fighter controller ordered 144 interceptors to deal with the imaginary threat confront-ing him from the Thames, and only half that number of aircraft to meet the real attack on Rouen. In the months that followed, 'Moonshine' was used on twenty-eight occa-sions in support of daylight attacks, with varying degrees of success: sometimes the Germans were misled, but sometimes they were not.

The difficulty was that one 'Moonshine' was required for each *Freya* to be covered, as the frequencies were differ-ent in each case. Since the German early-warning network was constantly expanding, more and more 'Moonshine' air-craft would have been needed. The most serious limitation was that the 'spoofing' forces had to remain out of sight of enemy ground defences and, since 'Moonshine' could only be used to support daylight attacks – night bombers did not fly in tight formation – this meant its use was limited to the support of shallow-penetration attacks. In the autumn of 1942, the Defiants flew their last 'Moonshine' sortie.

*

During the spring of 1942, British Intelligence improved its overall picture of the German night air defence system. In March, news arrived of an inland *Freya* radar station at Nieuwekerken; this was just to the north of the important German night-fighter airfield at Saint-Trond in Belgium. By this time, coastal *Freya* stations were known to be very frequent, but inland ones were very much a rarity. A high-flying reconnaissance Spitfire was sent to investigate. The photographs it brought back showed a *Freya* radar set and a cluster of searchlights, but the latter were grouped round a radar with a large circular bowl – like the one photographed in the Berlin Zoo. The radar's proximity to the airfield at Saint-Trond strongly suggested that this was some sort of night-fighter control centre. Shortly afterwards this was confirmed. An agent reported a night-fighter control centre at Domberg, on the Dutch island of Walcheren, and the subsequent high-level photographic reconnaissance revealed a *Freya* and two more of the 'Berlin Zoo' radar sets. A more detailed study of the original site in Belgium showed that there were in fact two 'Berlin Zoo' radars there.

This called for a closer look at the new type of radar. On 2nd May a reconnaissance Spitfire ran in fast and low, southwards along the Dutch coast. As it passed the Domberg site, the pilot started the sideways looking camera. Once again Flight-Lieutenant Tony Hill pulled off a low-level scoop: the cameras took excellent photographs of first one of the 'Berlin Zoo'-type dishes, and then the other.[4] As the sets were pointing in different directions at the time, he took two quite different pictures of the radar. Equally important, an operator had been about to climb the ladder to the cabin of the second set just as the Spitfire swept past. He stood watching, helpless, to become a human yardstick when the photographs were analysed.

One night two weeks later, a further ploy was adopted to elicit information about this radar's range: a Beaufighter aircraft flew in towards the Domberg area, closely watched by the British radar station on north Foreland. A German

4. Hill was killed a year later on a similar photo-reconnaissance mission near Le Creusot.

night-fighter rose to engage it, and a long and inconclusive combat followed. All the time, the RAF monitoring service was recording the instructions passed to the German pilot. In particular, they noted that he was not permitted to move more than forty miles from the radar station. This was a strong pointer to the maximum range of the German equipment.

The next major item of Intelligence came from a Belgian agent. He had managed to steal a map from a German headquarters showing the deployment of a whole searchlight regiment; as luck would have it, the map covered the area around Saint-Trond. The station at Nieuwekerken was marked by a lightning flash, and so were two other positions at Zonhoven and Jodoigne, some twenty miles on either side. Could these be fighter-control stations too? And if they were, was twenty miles the standard distance between adjacent sites? Photographs brought back by reconnaissance aircraft revealed that this was the case. By extrapolating the line, Jones and his staff were soon able to pick out five German night-fighter control stations strung out at intervals along an almost straight line.

During the summer of 1942, the clump of flags on the wall map in Jones' office sprouted shoots to either side of its original starting point in southern Belgium. The great radar hunt was on. Charles Frank christened the German defensive belt the 'Kammhuber line'. Agents were sent out to areas previously calculated to seek out the radar stations now identified as the Giant *Würzburg* sites. Not surprisingly, their catch was good: a gyrating paraboloid the size of a suburban house could hardly escape being the object of wonder and speculation by the local population. The descriptive vocabulary of the inhabitants of the low countries was seriously strained: 'inverted unbrella' and 'magic mirror' were typical of the terms used. It must be recalled that the ordinary man in the street had never heard of radar at this time. One Giant *Würzburg* was talked about so much that it became known as *'le fameux miroir d'Arsinont'*. RAF bombers dropped caged carrier-pigeons during their flights over Belgium, Holland and northern France; the birds' legs bore labels asking the finder to write in details of any rotating dish-like structures he might have seen in the area, and then release the pigeons. This method alone succeeded in locating three sites previously unknown to Dr Jones in the British Intelligence headquarters in London.

One point of controversy at this time was the system used by the Germans to control their searchlights. In general Bomber Command crews were of the opinion that the majority of the lights – in particular the hated blue-tinged ones – were radar-controlled in some way. During their 'ferret' flights, aircraft of No. 109 Squadron had several times noticed that the reception of 570-megacycle signals coincided with accurate exposure by searchlights. The difficulty was that none of Jones' many other sources had confirmed this view.[5] The photograph of the original Giant *Würzburg* near Saint-Trond had shown it to be surrounded by three searchlights, and from their intimate association it seemed reasonable to assume that the latter were directed by the former. It seemed to Jones to be a relatively simple matter to get agents to pinpoint a representative selection of searchlights, and then photograph them from the air to see whether there were Giant *Würzburg* radars associated with many of them. The pinpoints duly came in from the agents, and on 20th May a reconnaissance aircraft went out and photographed them all. When the photographs were analysed, the searchlight emplacements were found to be empty. By a quirk of fate, these had been removed under Hitler's general order of 5th May, stripping the Kammhuber line of its searchlights, in recognition of the improvements in its radar equipment position.[6] Jones concluded from the absence of searchlights

5. The truth was that the German system was a compromise between freely operating searchlights and strict radar control. The 1942 equipment for a searchlight battery normally comprised twelve lights, eight sound locators and one *Würzburg*. But many batteries did not have their *Würzburg* set up in the immediate area. The British system of radar controlled searchlights – which the German was expected to resemble – involved the use of one radar per searchlight.

6. See page 73.

that they and the radar sets belonged to two quite separate systems of fighter control, and that the latter system was in the process of replacing the former. (This had in fact happened six months earlier as far as the night-fighters were concerned; the German anti-aircraft guns were continuing to make use of the searchlights loosely controlled by radar.) The original photographs of the site near Saint-Trond must have been taken during the period of transition. On 13th July, Jones wrote:

> Many photographs have been examined and although *Würzburgs* have been identified with flak, none has yet been associated with searchlights except in one case (Saint-Trond) where three searchlights were grouped around a GCI [ground-controlled interception] *Würzburg*. They may have been used for signalling to night-fighters, but they were subsequently removed while the [Giant] *Würzburg* remained.
>
> It is therefore certain that most German searchlights are not radio controlled, and that even if the control exists it is not used on a scale sufficient to resolve the apparent conflict between 109 Squadron observations and information from all other sources.

Not until the autumn of 1942 was the matter of the German radar-controlled searchlights finally settled.

It might be asked whether it mattered whether the searchlights were found to be radar controlled or not. The fact was that there was – much later – an important corollary. Since the beginning of 1942, there had been a widely held belief among Bomber Command's aircrew that when they switched on their IFF ('identification, friend-or-foe') equipment German searchlights in the area tended to go out. There was only the flimsiest technical evidence to support such a theory. The IFF set was designed to receive pulses from British ground radar working on frequencies of about 25 megacycles, and reply to each with a coded pulse on the same frequency. It should not have transmitted at all without the initial pulse to trigger it; tests did, however, show that it sometimes transmitted a few pulses while settling down after being switched on. If the IFF had any effect at all, it could only have been on the radar set which was directing the searchlight; why the German operator should go so far as to douse his light, just because he had lost radar directions, could not be fathomed. Moreover the *Würzburg* worked on a frequency of 570 megacycles – a 25-megacycle signal could not possibly jam it directly.

In spite of this technical assessment, and Jones' belief that the searchlights were not radar controlled anyway, the supposition persisted that IFF interfered with enemy searchlights. It did no harm if the IFF was switched on when an aircraft was held by searchlights, and there was always the chance, however remote, that it might do some good. In June 1942 all IFF sets fitted to Bomber Command aircraft received a simple modification called the 'J' – or jamming – switch. This would enable the device to transmit continuously while the aircraft was over enemy territory. The modification was well-received, and crews reported that its use sometimes enabled them to escape from searchlights which held them.

Bomber Command's statistical staff carefully analysed the mass of evidence to see whether the innovation did in fact have any effect on losses. They found that it had no effect. During September, they issued a restricted-circulation report, which stated candidly:

There is no evidence that the use of the J-switch has had any appreciable effect on searchlights, flak defences or the activities of enemy fighters, or the 'missing' rate. It is known, however, that many crews think the device effective, and it should therefore be retained . . . Since no evidence has come to light indicating the harmful effects of the J-switch, the psychological effect on the crews alone is sufficient to justify its retention.

In retrospect, it is clear that where German searchlights did go out it was either because their carbon filaments had burned out or simply because they had detected no target within range; it was never due to any action on the part of the RAF crews. The bombers' ability to leave IFF switched on over enemy territories was to have important consequences on their losses later, however.

In the fourth week of October 1942, General Montgomery launched his great offensive at El Alamein, and within eleven days the Axis front had broken and the marathon retreat had begun. Now came the fruits of victory: Rommel's Afrika Korps had been well equipped with radar, and as the German troops withdrew many of the sets had to be left behind. Mr Derek Garrard flew out to Egypt to see what he could pick up. He soon found that the very quantity of material now available was likely to be an embarrassment, and when he returned to England he approached Mr J. B. Supper at the Royal Aircraft Establishment at Farnborough. Supper and his five-man team gladly undertook the work of examining the German radar equipment, but they could have had no idea of the magnitude of the task. Soon captured equipment was arriving at Farnborough by the lorryload, and Supper's men were aghast at what they found in the first few crates: the filth and the smell, the shattered state of the equipment, sizes and weights to be handled – all these made the task of analysis seem impossible. It would have to be handled as a major engineering project. Supper asked for large accommodation, mechanical handling equipment and five times as many men as he now had. Within two weeks the Director of the Establishment at Farnborough had emptied one of the precious hangars of its aircraft and made it available to Supper, and cranes, fork-lift trucks and men arrived to swell his team.

Garrard had laid down three main requirements for Supper's section: first, the collection of labels off the pieces of equipment from which valuable Intelligence could be gleaned; secondly, a general analysis of German radar techniques; and thirdly, a search for anti-jamming circuitry.

As Supper became more familiar with the German equipment, he felt inclined to attempt the full working reconstruction of the radar sets as well. A *Würzburg* set had been recovered almost intact. Its missing and broken parts were replaced with substitutes cannibalized from wrecked radar sets. To determine the circuit layout from the radar set itself took many hours of skilful and painstaking work – it is far more difficult than building a set from an existing circuit diagram. Supper's biggest worry was that some scarce component might be burned out by the premature operation of the set while not fully repaired: he had only one *Würzburg* transmitter valve, for example. To prevent this, each unit was assembled and tested independently, and studied for compatibility with the other units, before being passed as fit.

At last the captured *Würzburg* was ready for testing.

Supper recalls how they switched it on and it worked, though not for long, as it had its foibles and required some nursing. Gradually they learned to tame it, and soon they 'saw' their first aircraft on it: 'I telephoned the news to Garrard. Within a few hours he arrived at Farnborough accompanied by Jones and Frank. The three of them were like schoolboys on their first visit to the Science Museum. They jumped on and off the wooden platforms surrounding the radar, they turned handles and knobs, and they touched every part of it with loving caresses.' Supper was puzzled by this display of exuberance; he had naturally expected the Intelligence people to share his pleasure at a fine job of reconstruction, but why this boisterous joy? Gradually it dawned on him that for over two years these men had pored over maps and photographs, sifted and analysed reports from agents and slowly and painfully built up a mental picture of the enemy radar capability. Here for the first time was the tangible reality of their endeavours – a working *Würzburg* fully manned by an operational crew. And it had turned out to be exactly as they expected.

Supper's team was increased until it numbered some thirty-five men, including several RAF and USAAF personnel. Working *Freya* and *Seetakt* radars began to emerge from the rubble, and Supper was able to tell Garrard in some detail what to look for when he went 'beachcombing' on the battlefields of Africa, and later Europe, in order to make good his deficiencies. Garrard rarely let him down.

4

Towards the Offensive

> If you know the enemy and know yourself you need not
> fear the result of a hundred battles. If you know yourself but
> not the enemy, for every victory you will suffer a defeat. If
> you know neither you will always be beaten.
>
> – GENERAL SAN-TZU

As German troops advanced deeper and deeper into the
Soviet Union, it became clear in Britain that after almost a
year on the defensive, here was a chance to regain the initiative at last. Clearly an invasion of the continent was out of
the question; this would take time to develop, and it was
evident from the tide of German armed forces sweeping
eastwards that some means had to be found of striking at
the German homeland now. The only readily available
means was through Bomber Command of the RAF. During
the last weeks of the winter, Mr Churchill assured the Russian Premier that now that the season was improving, the
RAF was resuming the heavy air offensive against Germany: 'We are continuing to study other measures for
taking some of the weight off you.' The other measures were
not to materialize so rapidly, however, and in the middle of
April 1942, he admitted that the bombing of Germany was
'the only way in our power of helping Russia'.

In the meantime, Bomber Command had been through a
reorganization: the fact that the Germans had found it
necessary to develop radio aids for accurate navigation and
bombing led to some scepticism about the results that the
British bomber crews – who had no such aids – could be
achieving. Air Marshal Saundby recalls that soon after he

went to Bomber Command headquarters at the end of 1940, he told his staff that, as far as he could see, when a bomber force claimed to have dropped three hundred tons of bombs on a certain target, all they could be certain of was that they had 'exported three hundred tons of bombs in its direction'. A year later, Professor Lindemann had conducted his own investigation of the results of RAF bombing, and his findings were highly disquieting: of the crews who thought they had hit the target, only one in three had in fact placed his bombs within five miles of it; in the case of the Ruhr targets, the figure was as low as one crew in ten.

During September 1941, Mr Churchill taxed the Chief of Air Staff with this:

It is an awful thought that perhaps three-quarters of our bombs go astray . . . If we could make it half and half, we should virtually have doubled our bombing power.

As a long-term investment, the development of two radio devices to improve bombing accuracy was begun; these will be dealt with in later chapters.[1] The most urgent requirement was for some form of radio aid to improve the basic navigating accuracy of the RAF crews, however.

Fortunately work on such an aid was well advanced, and the set – called *GEE* – was already nearing the service trials stage. The *GEE* or 'Grid' system of navigation had been conceived by Mr R. J. Dippy back in 1938. At that time, no effort could be spared to develop it, and it was not until June 1940 that work on the device had begun in earnest. In its final form, *GEE* employed three ground transmitters each about one hundred miles from the other two. These transmitters acted in unison and radiated a complex train of pulses in a set order. In the receiving aircraft the navigator was provided with a special radar receiver, which enabled him to measure the time difference between the reception of the various signal pulses. By referring these differences to a special *GEE* map of Europe, the navigator was able to

1. These were to be code-named *H2S* and *Oboe*. See pages 137 to 140.

determine his position to within six miles, while flying up to four hundred miles away from the transmitter; at shorter range the accuracy was even better. In effect, the *GEE* transmitters laid an invisible radio grid across the continent; by listening to the character of the transmissions, a *GEE* operator could tell accurately on which 'lines' of the grid his aircraft was. Obviously this was a great improvement on *Knickebein*, the German system, for with *Knickebein* an accurate positional fix was possible only at the point at which the two Lorenz-type beams crossed.

The early history of *GEE* had brought a splendid example of what might be termed 'preventive Intelligence' work. The system began its service trials at the beginning of July 1941. The first experimental *GEE* sets had been fitted to Wellington bombers of No. 115 Squadron based at Marham in Norfolk, and first reports had been very favourable: the device had been tried out far over the North Sea, and the crews, who were unused to anything more complex than their own radio direction finders, reported the accuracy as 'uncanny'. Two aircraft used *GEE* during a bombing raid on a Ruhr town on 11th August, and all went well; but on the following night, after two *GEE*-equipped aircraft had attacked Hanover, one failed to return.

Only now was the great danger of these operational trials over enemy territory appreciated. History seemed to be repeating itself with a vengeance, for it was the similarly premature introduction of *Knickebein* which had resulted in its downfall almost a year before. Now the RAF was about to fall – if it had not already fallen – into exactly the same trap. It would be at least six months before the new *GEE* receivers became available in quantity. If the Germans were to learn the secrets of the new radio aid, they would certainly build up a jamming organization capable of blotting it out as soon as it was used on a large scale, just as No. 80 Wing had with the *Knickebein* beams. The jamming of radio aids was a game that two could play.

The Chief of Air Staff, Sir Charles Portal, ruled that the *GEE* trials should cease immediately. On 20th August, at Portal's request, Sir Henry Tizard held a meeting to discuss whether it was likely that the loss of the aircraft had in fact compromised *GEE*. The information was sparse: there had been no signals from the aircraft, and nobody had seen it go down. The *GEE* equipment had been fitted with ten detonators which could be set off by the aircraft's pilot, navigator or wireless operator, and these charges would have prevented the Germans from learning anything of use – provided they had been triggered off. But their very use would have alerted the German Intelligence services to the existence of something new and secret, and they would surely concentrate on trying to find out what. The consensus of the meeting was that there was a one-in-three chance that the Germans had found the missing *GEE* receiver, though probably in a damaged condition.

Dr R. V. Jones worked on the assumption that the Germans had got wind of the device, in which case they would soon be scratching around for clues much as he had been forced to do when unravelling the secret of the German beams in the summer of 1940. Even if the Germans had not picked up the missing *GEE* equipment, there were still several ways in which they could come across details of the new aid: they might learn about it from the indiscretions of RAF prisoners – seventy-eight Marham aircrew had been lost over enemy territory since *GEE* trials had begun. If the *GEE* had been too badly damaged to identify, the Germans would surely endeavour to find out what it was by examining its labels; was not that how he himself had always worked? It was possible in any case that the German listening service had picked up the *GEE* transmissions, or a German aircraft might have photographed a *GEE* transmitting station; or a German agent in Britain might have learned something about the new system.

Something obviously had to be done to save the situation. Even now, new aircraft were entering service with the mountings and cabling already installed for *GEE*, even though the special receivers were not yet available: if these fittings' labels were simply removed this alone would warn the Germans that this was connected with something new and secret. The best that Jones could hope to achieve was to mislead the Germans into believing that the new aid was something quite different from what *GEE* in fact was.

His solution was a masterpiece of Intelligence effort, and from German records we know that it did work. He set about removing all signs of *GEE* which might be of use to the enemy Intelligence service, and in its place he introduced a completely new navigational aid for the bombers, called 'J'. This was the nearest letter phonetically to *GEE*, so if the Germans eavesdropped on prisoners' conversations they could persuade themselves that they had misheard 'G' for 'J'. Having provided the real radio aid with an 'alias', all aircraft fitted with *GEE* were removed from the operational station at Marham and personnel were impressed with the need for secrecy while the Command waited for the equipping of its bombers with *GEE en masse*.

The genuine *GEE* transmissions were altered slightly so that these now looked like ordinary radar signals when analysed on a cathode-ray tube (to have switched the *GEE* transmitters off altogether after the loss of the 115 Squadron Wellington would have been too obvious a give-away). At each of the *GEE* transmitting stations additional aerials were erected so that their appearance, as well as their radiations, resembled those of the more common Chain Home radar stations. The type-number of the *GEE* set fell in the 3,000 range; and its '*R*' prefix gave it away as a pulse-type receiver, a fact that the Germans would probably have learned from other British radar devices captured (Jones hoped that the Germans would place as much reliance on the information to be gained from systematic equipment labelling as he did himself). Firstly, the incriminating type-number was removed from all aircraft cable and mounting installations; and then he substituted a new type-number, 'TR 1335', instead. This was a clever move, for the number in the 1,000 series and the TR prefix would strongly suggest to the enemy that the device was a communications transmitter-receiver, which they could then ignore.

Having done away with *GEE*, to all intents and purposes, it remained to create 'J'. Since the Germans had themselves

used Lorenz-type beams in two of their own radio navigational aids, they would not be surprised if the British followed suit. The J-beams were in short to be a copy of the
Knickebein beams. In the winter of 1941, three high-powered
Lorenz-type beam transmitters were accordingly erected in
eastern England to throw the appropriate beams over Germany. The beams went out on about 30 megacycles, which
the RAF bombers could pick up on their own Lorenz blind-
approach receivers. Dr Jones hoped that this slavish copying of *Knickebein* would flatter the Germans. To increase
the illusion, Bomber Command crews were encouraged to
use the J-beams when returning from their targets, as a
navigational aid. In a remarkably short time, the RAF's
navigators, who were then in no position to look such a
gift horse in the mouth, availed themselves of the new aid
to finding their way back home.[2]

Jones could not, of course, guarantee that these moves
would save *GEE*. Their success would be measurable only
as a delay before the eventual jamming of the device when
GEE finally became operational. It was too much to hope
that the Germans would continue to be misled long once
Bomber Command started to use *GEE* on a large scale
over enemy territory. The most sanguine estimate was that
GEE would have an unjammed life of about three months
once it re-entered service; and then the Germans would
find out how it worked and set up a suitable jamming organization. British Intelligence had no way of learning
whether the Germans had already got that far or not, until
the first large attacks using *GEE* began.

By the beginning of March 1942, sufficient *GEE* receivers
existed to equip thirty per cent of Bomber Command's air-

2. Thus a February 1944 German Intelligence report on British
navigation techniques described: 'There are various movable beam
transmitters (Jay beams) in GB rather like the German *Knickebein*
transmitters, for marking turning points on the outward route and
to prevent crews losing their way on the return flight.' The system
was 'seldom used' as the British were said to fear German countermeasures.

craft. Whether the effort to get the system operational again would prove to be a complete waste as a result of the previous year's indiscretions would soon be known. From the 8th March onwards, *GEE* receivers were used in large numbers over Germany again. The months passed, and the system remained unjammed, so it seemed that the efforts to disguise *GEE* had paid off well. By now there could be no doubts but that the Germans did know about *GEE*, for some twenty *GEE* aircraft had been lost during the first month of operations alone. But June and July came and went, and still the system remained undisturbed, much to everyone's surprise. Night after night, the British bombers went out over Germany in force, determined to wring the utmost out of the condemned device, affectionately known as the 'Goon box' to the crews.

The first captured *GEE* sets had indeed aroused considerable interest in Germany during March. The principle was not new to the German air force, as Telefunken had been developing a similar device in 1939, but had to cease work on it when Hitler had ordered research on such long term projects to cease at the end of 1940. The first *GEE* receiver to fall into their hands had been taken from the wreckage of a Wellington which had come down in the sea off Wilhelmshaven on 29th March. Although the set had been damaged by salt water, it was the water that had saved it from destruction by the incendiary charges fitted to its casing; as they had abandoned the Wellington, the crew had initiated the detonator system, but there was a built-in delay to give everyone time to get clear and the water had smothered the charges.

The procedure adopted by German Intelligence from this point on was just as Jones had imagined it would be. Intelligence and radio officers hunted carefully for any further scraps of information relating to this newly discovered British system. Engineer-Colonel Schwenke, the Intelligence officer in charge of the air force section dealing with captured enemy equipment, discussed the find at a Berlin conference on 26th May. He explained that the remains of this equipment had been recognized in several shot-down British aircraft, but in every case except the Wilhelmshaven incident and a second aircraft which had also crashed in the sea, the units had been smashed beyond repair or reduced to molten metal by their demolition charges.

Schwenke continued:

We have also carried out a systematic interrogation of prisoners. The following facts have come to light. As a result of the extensive use by us of the *Knickebein* and *X*- and *Y*-*Gerät* systems these devices fell into British hands; this was because we did not fit demolition charges.

In mid-1940 orders were given for the immediate construction of copies of the *Knickebein* and a year later, in August or September 1941, these were ready for service. The British found it comparatively simple to copy the German set, as the airborne *Knickebein* uses the installation for [Lorenz] blind beam [airfield approach] receivers, and the British had obtained the licence for the Lorenz set before the war . . . From the interrogation of prisoners, we know that this system was used under the designation 'Julius'.

'Julius' was the letter 'J' in the German phonetic alphabet. All in all, it was obvious that Dr Jones' 'plant' had succeeded exceedingly well, as the Germans suspected nothing.

Having discussed the mysterious 'Julius' system, Colonel Schwenke returned to the system revealed by the captured sets. 'It has also been established,' he explained, 'that the British gave orders at the same time for the development of a new system which gives the pilot his position at all times. The equipment for this is the receiver I have just described. Tests have been carried out on it by Telefunken, but the set was unfortunately not received in good condition. Our experts are still not in complete agreement concerning the technical workings of the equipment.' He described accurately the manner in which *GEE* was employed, and the principle on which it worked, and added that the equipment was being installed as standard in the RAF's principal bomber types – Wellington, Lancaster and Halifax: 'I think it is being used not so much to find pinpoint targets as to improve dead-reckoning navigation.'

The plan was to install the captured equipment in a German aircraft and establish its actual degree of accuracy by picking up the British *GEE* grid transmissions. The transmitters currently in use had become known to the Germans 'through captured material'. They were located on the south-eastern coast of England in positions where they could cover the Ruhr district. Again confusing *GEE* with the camouflage '*J*', Colonel Schwenke talked of the three transmitters' beams 'converging approximately over Dortmund'. The possibility of jamming was being investigated, but first it was necessary to find out exactly how the system worked and its frequencies. The ground transmissions were fairly powerful; they could be jammed if more powerful jamming transmitters were switched on and operated on the same frequency, but 'it is not all that simple'. Schwenke told the conference that General Martini was calling a conference on the question of jamming *GEE* shortly afterwards.

The Germans were quick to appreciate the significance of *GEE*. Colonel Pusch took charge of the programme of countermeasures, and by the end of July 1942 a special unit had been formed to organize the jamming of *GEE* – the Second Battalion of the Air Signals Radio Monitoring Regiment (West). Dr Mögel, a German Post Office technician, converted a number of speech transmitters into makeshift jammers and these were installed round important targets in Germany. On 4th August 1942, the blow so long expected by the RAF finally fell, and the jamming began. That night, twenty-two *GEE*-equipped aircraft leading a small attack on Essen in the Ruhr reported the system's signals were swamped by enemy interference when they were still some twenty miles from the target.

Within a short time, these improvised jamming transmitters gave way to more powerful jamming equipment specially designed to counter *GEE*, under the code-name *Heinrich*. The transmitters were set up all over German-occupied Europe, and one even found its way to the top of the Eiffel

Tower in Paris. Within the next three months so many jammers were brought into operation that *GEE* was rendered unusable over the whole of Germany and western Europe. After a time, the RAF came to rely on *GEE* mainly as an aid for navigation when clear of enemy territory – the same role to which the German air force had relegated *Knickebein* after the autumn of 1940.

The first measures counter to the German 'Kammhuber line' did not involve jamming at all; during the first two years of the RAF Bomber Command's night offensive, the bombers had invariably made their own way to their targets. There was no attempt to keep the force together – indeed many crews attributed their survival to the fact that they did something different from everybody else. A large number of bombers crossing the 'line' at widely separated points over a period of several hours – this was the grist for which Kammhuber had designed his mill. But the standard unit of the 'line' was its 'box', patrolled by one radar-guided fighter aircraft; so the one great weakness which was realized in England the moment that the line's workings became clear was that it could be easily overwhelmed at any one point. Each defensive box could only engage one bomber at a time, and during each engagement – an average of ten minutes – there was an unguarded gap in the line. If the raiders were to fly in a tight mass all would pass through the line unscathed bar the one or two unfortunates upon whom the German fighter controllers focused their attention.

Hitherto the bombers' navigation had not been precise enough to enable the bombers to be 'streamed'; the advent of *GEE* had changed all that. The concentration-in-time of bombing attacks promised a further advantage, for it would reduce the effectiveness of the German anti-aircraft guns: the target's defences would be saturated.

At the beginning of the year, Dr B. G. Dickins, head of the Bomber Command operational research section, wrote:

Both *en route* over land to the target and in the target

area, concentration in time and space together with dispersion in height, will result in the minimum of aircraft
casualties, by day and night.

In a given area and time, the number of aircraft that
can be engaged by searchlights and/or guns is limited.
Thus, by passing concentrated streams of aircraft over
a given route, the enemy will be presented with a difficult
problem owing to super-saturation of his defences, so that
our losses should be small. By passing a concentrated
stream of aircraft at different heights over the area, the
enemy will experience great difficulty in determining the
exact course, and particularly the height, of individual
aircraft. Should he resort to barrage fire, the dispersion in
height will render this particularly ineffective.

The new tactics were first tried on the night of 30th May
1942, during the famous Thousand Bomber Raid on
Cologne. The attacking aircraft all took the same route,
and the period of attack was cut from about seven hours
to two and a half – an average of seven aircraft bombing each
minute. Forty-one bombers – only 3·8 per cent of the force
– failed to return. This was a significantly lower proportion
than had been lost up to this time. The 'bomber stream'
had been born.

Despite this success, the bunching of aircraft at night was
a source of argument in the Command for some time. The
new tactic was unpopular with the crews: several crews
returned to base with holes in their wings caused by incendiary bombs dropped from above. Dr Dickins had to reduce
the risks to a simple mathematical argument, by estimating
what the actual chances of collision were. 'It became quite
obvious to us that while a collision was something like a
half a per cent risk, the chance of being shot down by flak
or fighters was a three or four per cent risk,' Dickins later
recalled. 'So we could allow the collision risk to mount
quite a bit, provided we were bringing down the losses due
to other causes.'

Kammhuber's answer to Bomber Command's new tactic
was the buttressing of his line with additional stations in

front of and behind it, thus increasing the depth of the defences the raiders had to penetrate. In the autumn of 1942 British losses started to rise again.

In the meantime, Bomber Command's picture of the German night defensive system was complete in every major detail except one – there was little information about the interception device fitted to the German night fighters. During the spring of 1942, radio monitoring stations in England had first noted a new code-word in the German night-fighter crews' vocabulary – '*Emil-Emil*'. From the intercepted radio conversations, it became clear that *Emil-Emil* must be some form of target location device carried in the aircraft. On 23rd July, two months after the first thousand-bomber raid, a German night-fighter was heard to radio : 'I have got the enemy aircraft on *Emil-Emil*. Please give further vectors.' And again on 6th September, a ground station was heard to ask a fighter aircraft: 'I have vectored you to within two kilometres of the enemy aircraft – haven't you picked him up on your *Emil-Emil*?' Finally, on the same night a fighter aircraft was heard making an explanation of breaking off radio contact with ground-control: 'I had got an enemy aircraft on *Emil-Emil* and broke off radio contact with you during that time.' The ground controller rebuked him and told him that he had to remain in radio contact at all times.

By October 1942 the volume of radio traffic with explicit references to *Emil-Emil* had reached such proportions that Jones concluded that the entire night-fighter force either had, or was about to get, the device. The device was clearly very important to the Germans, but what sort of device was it? Almost certainly it was some form of radar, or an infra-red homer which detected the hot exhausts of the British bomber aircraft. It was vital to know which of these two principles had been employed, and if it was radar, the frequency on which it worked.

The TRE scientists at Malvern therefore set up a special monitoring station on the Norfolk coast: the radar receiver was sited in a ground depression so that – barring the most abnormal radio propagation – it was below the normal horizon and would pick up only those transmissions emanating from aircraft. Within a few days the men had what they were looking for: a cinematograph camera mounted in front of the cathode-ray tube photographed a series of pulses on a frequency of 490 megacycles. This was part of the radio 'spectrum' not used by the RAF, and the rate at which the signals' sources changed their bearing was consistent with the notion that they must be fast-moving aircraft.

In the days that followed, scores of similar signals were picked up. But the hunt was not over, for there was no proof the enemy aircraft concerned were night-fighters: they might for example be coastal patrol aircraft, installed with special radar for searching for ships. The only way to be absolutely certain was to send out 'ferret' aircraft to trail their coats in front of the enemy.

It was outside Dr R. V. Jones' power to hazard aircraft and men's lives in this way. He asked for Cabinet approval for the necessary sorties. This was immediately given, and the Prime Minister himself demanded the most vigorous action to settle the mystery of *Emil-Emil*.

The unenviable task of acting as live bait for the German night-fighters fell to No. 1473 (Wireless-Investigation) Flight of the RAF. The unit began sending out 'ferret' aircraft over France, Belgium and Holland. In England radar stations kept careful watch on them, ready to flash immediate warning should they see night-fighters preparing to close in. The Germans ignored these isolated aircraft, and although the 'ferrets' did record suspicious signals around 490 megacycles the mystery remained unresolved.

Since the Germans would not react to listening-aircraft in the coastal areas, it was obvious that the aircraft would have to accompany the actual bomber formations to their targets; they would certainly not be ignored there. This was a distinctly more hazardous duty for the listening-aircraft. Shortly after 2 a.m. on 3rd December, a Wellington aircraft – DV-819 – climbed into the cold night air from its airfield at Gransden Lodge near Huntingdon, ordered to accompany the bombers attacking Frankfurt that night. The Wellington's crew was directed to listen for radar-type transmissions on frequencies round 490 megacycles. To ensure that they came from a German night-fighter's radar, they were to allow the enemy to close in to attack, following his radar transmissions all the time; then the British crew were to radio their findings back to England and escape as best they could. Rarely even in war have men been asked to undertake a mission with a slimmer chance of survival.

By half past four, the Wellington was ten miles west of Mainz and its captain, Pilot-Officer Paulton, turned northwards on to the first leg of the homeward journey. One minute later his special radio operator, Pilot-Officer Jordan, picked up weak signals on 490 megacycles. This was what he had been briefed to look for. For the next ten minutes he watched intently, noting their characteristics and observing that they were increasing in strength. Jordan warned the rest of the Wellington's crew of what was happening, and he drafted a coded signal to the effect that 490-megacycles signals had been picked up, almost certainly emanating from a night-fighter's radar. He passed the message to the wireless operator, Flight-Sergeant Bigoray, for him to transmit to England.

Any last doubts that the signals were from a night-fighter were shortly dispelled. They rose to a level which completely swamped his receiver, and he shouted to the crew to expect an attack at any moment. Almost immediately the Wellington shuddered under the impact of exploding cannon-shells. Its pilot threw the aircraft into a steep diving turn to shake off their assailant, and the rear gunner opened fire on the attacker, which he recognized as a Junkers 88. Jordan had been hit in the arm, but he wrote out a second coded message confirming that the frequency given in his first report was without doubt that of a night-fighter's radar. The rear-gunner fired about a thousand rounds at the attacker, but then his turret was put out of action and he himself was hit in the shoulder. The German pilot came in for attack after attack, and the special radio operator received further wounds in his jaw and eye. Then the night-fighter vanished, leaving the Wellington aircraft a shambles and barely flying. The port engine's throttle had been shot away, and the starboard throttle was jammed. Both engines were spluttering, the two gun turrets had been knocked out and the hydraulic system was shot to pieces. The starboard aileron no longer worked, and both the airspeed indicators were useless. Four of the crew of six were badly wounded.

Despite injuries suffered when one of the German's cannon-shells had exploded near his legs, Bigoray tapped out Jordan's second coded message. He could get no acknowledgment, but he repeated the message again and again in the hope that someone might hear. At five minutes to five a ground station in Britain did pick up the vital signal, and radioed a reply, but the reply was not heard because the Wellington's receiver was damaged. Although the crew's sacrifices had not been in vain, they had no way of knowing it. Bigoray continued doggedly tapping at his Morse key as the Wellington crossed the French coast near Dunkirk at a quarter to seven in the morning.

Half an hour later, the battered aircraft struggled over the coast of England. But now there was a new problem: the aircraft had been too badly damaged to risk a crash landing, so Paulton decided to bring it down in the sea near to the shore.

One of Bigoray's legs had stiffened from its wounds, and they knew that he might not be able to get out of the aircraft before it sank. The only way was for him to parachute out. The wireless operator laboriously hauled himself to the rear escape hatch, only to remember that he had not locked down his Morse key, to provide a continuous note signal to enable the ground direction-finding stations to track the stricken aircraft on. Bigoray crawled back and performed his final duty as the Wellington's wireless operator. He parachuted out of the Wellington over Ramsgate and landed safely, with a copy of the second vital signal in his pocket.

Paulton ditched the aircraft in the sea some two hundred yards off the coast at Deal. The crew inflated the rubber dinghy, but it was holed many times and quite useless. The men floundered back on to the wallowing bomber. A few minutes later a small boat appeared, and took them ashore.

All too often, acts of extreme bravery by individuals have no real effect on the outcome of a war. This act did. On 29th December, Jordan was awarded the Distinguished Service Order, Paulton the Distinguished Flying Cross, and Bigoray, the wireless operator, the Distinguished Flying Medal. By a combination of boldness, imagination, patience and good fortune, British Intelligence had built up a most detailed picture of the German night defensive system; with few omissions, they were to detect every evolution and change in the methods used by the German air force during the months that followed. Now the vulnerability of the radio devices used by the Germans could be exploited to the full.

5

Doubts and Decisions

We must be perfectly clear about the fact that the enemy is making every effort to outstrip us in the field of high frequency technique and its operational employment. We must, therefore, pit all our technical and productive power against him, so as not to allow his tricks to become effective.

– COLONEL-ENGINEER DR MÖGEL, during a lecture to German Air Force signals specialists in February 1943

The RAF's first attempt to defeat the German early-warning radar had centred on a 'spoof' device, 'Moonshine'; but this had been limited to supporting daylight attacks. What was really needed was something to counter the mounting opposition to the British night attacks. Dr Cockburn's team at the Telecommunications Research Establishment produced a jammer to obliterate, as distinct from deceive, the signals from the *Freya* radar system. The new device, 'Mandrel', had the same effect on the enemy radar as a car with an unscreened ignition has on domestic television. This type of interference, the crudest form of jamming, was known as 'noise jamming'. The new equipment transmitted a spread of noise to cover the frequencies used by *Freya*, from 118 to 128 megacycles. 'Mandrel' was rushed into production during the summer of 1942.

RAF Bomber Command would also have liked some means of noise-jamming *Würzburg*, but the system was a very difficult target for this. It was worked on too high a frequency for enough power to be generated to swamp the echo signals.

By the end of November, enough 'Mandrels' had been

built for the device to be used in action. The plan was to jam the *Freya* early-warning radar chain first, and then the inland *Freya* sets used for tracking the narrow-beam Giant *Würzburg* radars round on to their targets. The former would be dealt with by nine 'Mandrels' carried in nine Defiants of No. 515 Squadron, ordered to orbit set positions spaced out along a 200-mile line fifty miles off the enemy coast. When the jammers were switched on, a two-hundred-mile stretch of the German early-warning system would be 'taken out' by this 'Mandrel' screen. To deal with the enemy's inland *Freya* system, two bombers in each main force squadron were also fitted with 'Mandrel' sets; it was anticipated that these aircraft, distributed throughout the bomber stream, would defeat the working of Kammhuber's fighter-control system.

Each bomber aircraft was also fitted with a second noise jammer, aimed at the ground-to-air radio communications of the German night-fighter system.[1] Once over enemy territory, the RAF wireless operator was to 'sweep' his receiver over a certain waveband until he found any radio transmissions not in English; he was then to tune his own aircraft R/T transmitter to the same frequency and switch on a microphone installed in the extremely noisy engine-bay of his bomber. Thus he jammed all the German fighters' communications with a flood of engine noises.

All these developments left the *Würzburg* system still unjammed. This was not to say that there had been no consideration of such a countermeasure, for this was the most controversial episode of all.

By far the simplest method of confusing a radar picture is that of cluttering up the screen by means of metallic objects dropped from aircraft. As with radar itself, this form of countermeasure had been thought out in both England and Germany. The tactic was to highlight the differences which existed in the direction of the scientific effort of both countries. In England, Mr Churchill's scientific adviser Lord Cherwell (Prof. Lindemann) was able to use his great personal influence to the full during the important discussions

1. Its code-name was 'Tinsel'.

surrounding this tactic's introduction. Since the war he has been criticized for the way in which he used his influence, but when it came to major issues it is clear that his judgment was generally sound.

The principle of jamming radar by dropping clouds of metal strips had been mooted some time before war broke out. In June 1937, Dr R. V. Jones had visited the research establishment at Bawdsey, and here he had been impressed by the great sensitivity of radar, which was a rival system to the infra-red aircraft-detection device on which he was himself working. Radar could detect a single length of wire one-half of a wavelength long, a 'dipole' suspended beneath a balloon, at a range of twenty miles.

Jones later recalled: 'I thought about the implications of this, and in the following September or October Lindemann came into my room in the Clarendon and said something like this: "I heard from Winston that they are trying to get infra-red stopped." The *they*, as I understood it, referred to Tizard and Watson Watt. In the ensuing conversation, I remarked that this might be a pity since although infra-red had its weak points, radar was vulnerable too. All that you had to do was to sow a field of dipoles over the North Sea and the radar screens would be full of echoes. Trying to detect an aircraft through this would be like trying to see through a smokescreen. Lindemann said: "That's a good idea – I will get Winston to raise it." '

The 1938 draft of Professor Lindemann's thoughts on the matter still exists, and this is almost certainly the text of what he put to Mr Churchill:

Lest too much reliance be placed upon the RDF methods, it is perhaps worth pointing out that certain difficulties may easily be encountered in actual use. Though undoubtedly excellent for detecting single aircraft or squadrons thereof, flying together, it seems likely that great difficulties may be encountered when large numbers of aeroplanes attacking and defending are simultaneously in the air, each sending back its signals. This difficulty may be very materially increased if the enemy chooses to blind the RDF operator by strewing numbers of oscillators in the appropriate region. Such oscillators need consist merely of thin wires fifty to a hundred feet long, which could easily be suspended in suitable positions from toy balloons or even, if only required for half an hour or so, from small parachutes. As far as the RDF detector is concerned, each one would return an echo just like an aeroplane.

As Professor Lindemann added, it would be possible to distinguish an aircraft from single 'oscillators' after a short time since the one moved and the other did not; but if *thousands* of wires were dropped, it would be almost impossible to sort out the aircraft from the false echoes. He concluded: 'It is known that a system similar to the RDF was patented in Germany a good many years ago,[2] and it seems unlikely that the enormous volume of signals of peculiar type emitted by our station should not have been observed and interpreted by the German scientists. In these circumstances it is to be apprehended that they would adopt some simple counter method such as has been outlined with disastrous results if we have failed to take this into account.'

At the time, Mr Churchill was a member of the Air Defence Research Sub-Committee. Some weeks later, Jones asked Lindemann if Mr Churchill had raised the matter there, and he was told that, while Tizard and Watson Watt had admitted to Churchill that such jamming might have some effect, it did not look as though they were going to do any trials.

There the matter rested until 1940. The only documented evidence that the idea occurred to others in the interval was in the record of a conference at the Telecommunications Research Establishment, where it was stated that if friendly aircraft dropped metal strips at pre-set rates, they could be identified on radar as friendly. Professor Lindemann had resurrected the idea of using reflective metal strips in the

2. This would seem to have been a reference to Hülsmeyer's 1904 German patent for a 'Telemobiloscope'.

middle of 1940, but this time it was as a possible means of pushing the German beams off target by dropping the strips to one side of them. Dr Cockburn of TRE proved with a small sum 'on the back of an envelope' that it wasn't possible: to have collected enough energy for the system to be worth while, the strips would have had to be released in the centre of the enemy beam, but there they would not have deflected it at all.

In the summer of 1941, the RAF actually tried the idea of dropping metal strips to confuse enemy radar, in action in North Africa. During September a Wellington of No. 148 Squadron had been making a series of radio-investigation flights over the German positions, carrying a special wireless receiver which necessitated an unusual aerial arrangement. To their surprise, this crew found themselves almost always singled out as the main target by the German AA guns, even when other aircraft were about. Perhaps the aerials caused an exaggerated echo on the German radar? During their next attack on Benghazi, No. 148 Squadron's Wellingtons all accordingly dropped packets of aluminium strips eighteen inches long and an inch wide – the size of the aerial elements fitted to the original investigation aircraft. In the event, the strips achieved nothing, probably because these particular guns were controlled by sound locators. Perhaps the wind rushing through the original aircraft's aerials had increased its volume of sound. The strips were used only on one other occasion, and then the idea died a natural death.

The idea was resurrected soon after, and the Telecommunications Research Establishment began trials using metal strips as a countermeasure to radar. And it got a name: according to Mr A. P. Rowe, the establishment's wartime superintendent, Dr Cockburn called on him one day to discuss a suitable code-name for it, and Rowe required that essentially it should bear no relation at all to the device it covered: 'I looked around the room. "Why not call it something like 'Window'?" I said.' And 'Window' it became.

Much thought was devoted to concealing the true purpose of the 'Window' which would fall into German hands. One idea was to sandwich the metal foil between two sheets of paper, on each of which would be printed a tract calculated to improve the German mind. The War Office Political Intelligence Department provided information on the sizes of the current British propaganda leaflets, and Mrs Joan Curran, the only lady scientist at the TRE, was asked to experiment along the lines suggested. During the initial trials she was limited to the use of foils of about $8\frac{1}{2}'' \times 5\frac{1}{2}''$ and $8\frac{1}{2}'' \times 11''$ in size, the size of the PID's propaganda leaflets. She began by endeavouring to find the best shape for the metal foil, in other words the shape which would produce the largest radar echo, for the lightest amount of foil. A copper foil was used at first, cut into various forms, and strings of reflectors were fixed together like streamers in a kite's tail. Others were cut to resemble a step-ladder. It soon became clear that such complications were unnecessary – the simple oblong gave the best results. Mrs Curran found that the best material was the kind of tinfoil used in the manufacture of radio-condensers.

By March 1942, her initial technical investigation was complete. Seventeen flights had been made, during which the effect of 'Window' had been observed on every type of British radar set using a frequency of 200 megacycles or greater.

Mrs Curran reported on 22nd March:

It has been demonstrated that for frequencies of the order of 200 megacycles or more, echoes can be produced by jettisoning material from an aircraft, and that the quantities of material necessary to give rise to an echo equal in magnitude to that of the aircraft are not in any way excessive.

A bundle of 240 of the smaller sheets produced an echo approximating to that from a Blenheim bomber, she reported; the effect lasted for about fifteen minutes, if the bundle had been released at about 10,000 feet. Ten such

'Window' clouds over one mile would saturate the radar
screen and make it virtually impossible to pick out the
echoes from aircraft in the area. Most important was her
finding that the radar which suffered most was the new
British Type II ground-radar; this used the same frequency
as the German *Würzburg*.

The gist of Mrs Curran's report was quickly appreciated
by the Chief of Air Staff, and on 4th April he ruled that
Bomber Command could start using 'Window' to support
its operations as soon as the preparations were complete.
The Command's operational research section developed
suitable tactics for use with the tinfoil strips, and decided
that since each aircraft could carry only a limited quantity,
and the tinfoil itself was scarce, its use should be confined
to the target area. The Vanesta company received a large
order for the manufacture of the necessary strips. It was
now realized that the Germans would quickly discover the
cause of the jamming, if the tactic was as successful as the
TRE tests indicated, so there was little point in camouflaging
the strips as leaflets. By dint of much hard work, Vanesta
delivered a first consignment of 'Window' to Bomber Command early in March 1942, in time for the thousand-bomber
raid on Cologne; but the consignment arrived just after the
portcullis slammed down. Sir Charles Portal, the Chief of
Air Staff, had withdrawn his permission to Bomber Command to start employing the countermeasure as soon as they
were ready.

The reversal had been brought about by Sir Sholto
Douglas, Commander-in-Chief of Fighter Command. He felt
that in spite of the implications of the trials already held, his
own force would be poorly placed if the Germans chose to
employ this form of countermeasure against the British.
He prevailed upon both Portal and Lord Cherwell (Prof.
Lindemann) to delay its introduction until there had been
a more comprehensive series of trials against the whole
range of British radar sets. Portal discussed the matter with
his Secretary of State and with Sir Arthur Harris, the Commander-in-Chief of Bomber Command, and then ruled
on 5th May that the use of 'Window' was to be postponed

indefinitely pending the trials for which Fighter Command were asking.

On Lord Cherwell's insistence, this second series of 'Window' trials was conducted by one of his protégés, Dr D. A. Jackson. Derek Jackson was one of the more colourful characters in the history of radio-countermeasures: having studied under the then Professor Lindemann at Oxford, he had himself become a don and an acknowledged expert on spectroscopy. He had not lacked money – his father owned the *News of the World* – and he had enjoyed life to the full. In 1935 he had entered and ridden his own horse in the Grand National.

When war came, he had joined the RAF where he had rapidly gained the DFC and a reputation as an extremely skilful night-fighter radar operator. Prior to his new appointment he had been serving as Flight-Lieutenant at Headquarters, Fighter Command, which hardly seemed a suitable post for a man of his talents. It was the work with 'Window' that was to make his name at the Telecommunications Research Establishment. His great asset was an ability to bridge the difficult gap between the scientist and the operational air-crewman.

D. A. Jackson used the airfield at Coltishall near Norwich as a base for the 'Window' trials. The local night-fighter unit, No. 68 Squadron commanded by Wing-Commander Max Aitken, afforded considerable assistance. Within a few days the first results were available, and they were most ominous: the RAF's own centimetric-wavelength radar for night-fighters, AI Mark VII, suffered severely from 'Window', and the much older Mark IV, which used a frequency of 200 megacycles, was affected, but to a lesser degree. Sir Sholto Douglas asked the Air Ministry to prohibit Bomber Command from using 'Window' until the RAF had developed an antidote, in case the German Air Force should launch heavy air attacks using the tactic themselves. Sir Charles Portal agreed.

The Official Historians wrote that 'this surprising decision had allowed the threat of the much inferior and diminishing

German bomber force to deny an important tactical advantage to the much greater and increasing striking power of Bomber Command', but this seems an ill-considered judgment: at the beginning of April 1942, the German air force had 578 medium bombers available for operations. During May 1942 the average strength of RAF Bomber Command, on the other hand, was 417 aircraft; of these 136 were four-engined heavies, but 71 were Blenheims, Whitleys and Hampdens of little value and about to be retired. There was no guarantee that Russia would survive the German military

BUCK RYAN

A Military Intelligence officer drives Buck Ryan and Zola from the late Lord Brompton's shooting estate to Ack Ack Headquarters, where Ryan now reports his findings

···THOSE ARE THE SPOTS FROM WHERE THESE NAZIS INTENDED TO WORK. THEIR SCHEME WAS SIMPLE AND BASED ON THE ELEMENT OF SURPRISE. THEY HOPED—

TO DISORGANISE OUR RADIO LOCATION CONTROL AT SUCH TIME·WHEN THE ENEMY DECIDE TO INVADE OUR COUNTRY. IT WOULD BE AT NIGHT TIME AND NAZI AGENTS WERE TO TOW BOX-KITES—MADE WITH METAL ALLOY FRAMES—FLYING AT ABOUT 5,000 FT AROUND THAT ZONE OF ACK-ACK-GUN SITES. CARS WERE TO HAVE BEEN USED FOR TOWING—HIRED OR STOLEN!

advances of mid-1942, and if she had succumbed the whole of the German bomber force would have returned to the West. The German bomber force would have carried a far greater load over the shorter distance to the British capital than RAF aircraft could then have carried to Berlin. On this evidence, it is hard to challenge Portal's decision in early May 1942.

Lord Cherwell supported Portal's stand, and towards the end of the month he explained why in a minute to the Vice-Chief of Air Staff, Sir Wilfred Freeman. In brief, he did not believe that on present showing 'Window' could bring about a substantial reduction in Bomber Command's losses, currently averaging 6·5 per cent of aircraft dispatched. Of this

6·5 per cent, he attributed only 0·5 per cent to fire from AA guns working without searchlights, mostly in barrage; and of this 0·5 per cent he would attribute at most one-fifth to radar-controlled gunfire. In short, one aircraft in every thousand aircraft dispatched would be saved by the use of 'Window' over Germany, as things were at present. This, he suggested, 'is a poor justification for the premature release to the enemy of a device which is more effective against all our newer RDF than against his, and to which we have no effective reply in sight'. While he felt it right that pre-

parations for its use should go ahead, he believed that until bomber losses rose significantly, or there was an exceptionally difficult major operation where such a countermeasure might just tip the scales, or the British themselves developed an antidote to 'Window', it should not be used.

With the benefit of hindsight, it is easy to see that Lord Cherwell's arguments were based on wrong premises. But in May 1942, the layout of the German air defences was not fully known in England; it was still not fully appreciated that the German night-fighters, and not only the anti-aircraft guns, relied heavily on the *Würzburg* radars which were to suffer most from 'Window'.

*

Lord Cherwell discussed the 'Window' dilemma with Air Commodore Lywood, head of the RAF signals organization, on 18th June. Cherwell said he would withdraw his objections, provided that some form of self-destroying tinfoil strip could be produced; but even so, it was realized that the moment a bomber crashed with undestroyed 'Window' aboard, the secret would be out, and secrecy was the prime consideration.

Ironically, even as Cherwell and Lywood were in conference, news-stands all over England were selling a reasonably good working description of the 'Window' principle. In that morning's edition of the *Daily Mirror* the cartoon character, Mr Buck Ryan – a sort of moral James Bond – had again saved Britain from a terrible fate: German agents had planned to neutralize the British 'radiolocation devices' by towing metal-framed box-kites around the anti-aircraft sites.[3] While the baffled gunners tried to sort things out, the German troop-carrying aircraft were to have sneaked in and landed their men. As Mr Ryan finished his explanation of this latest piece of Nazi devilment, an army 'Intelligence officer' congratulated him: 'Yes, and it might have worked twelve months ago, Ryan. But not today!' How wrong he was. The next day Lywood sent a clipping of the offending cartoon to Lord Cherwell, with a covering letter:

I don't know whether the Hun is a subscriber to the *Daily Mirror*. If, as I suspect, he is, the attached cutting seems a fairly good investment in basic ideas.

Cherwell was informed that security officers were taking the matter up. If they did so, they did so subtly; the cartoonstrip's creator, Mr Jack Monk, recalls nothing of the incident. The fact that he had no scientific qualifications whatever serves to underline the basic simplicity of the 'Window' idea, however. He had thought the matter up unaided, and had no inkling of its wider implications. Soon

3. The fact that British forces used 'radio location devices' had been announced in 1941.

afterwards, the *Daily Mirror* was required to send engravers' proofs of all cartoons to the Ministry of Information for censorship. No one could think why at the time.

While D. A. Jackson worked out the best tactics for use with and against 'Window', others were doing much the same thing – in Germany. How similar was the early history here! The earliest experiments were carried out on the Düppel estate near Berlin, and the German air force's equivalent of 'Window' became known as '*Düppel*'. During 1942, aircraft conducted trials over the Baltic, dropping them in showers while German radar sets of various types were used to observe the result. The German scientists gained the same impression as their British counterparts – this countermeasure was dynamite. If the British were to find out about *Düppel* through some German indiscretion, it would spell the end of the German air defence system. The Air Force's technical office passed the results of the tests to General Martini, head of the signals organization. Martini presented a two-page report on the subject to Reichsmarschall Göring, stressing the grave danger should the RAF ever use the strips.

Göring was horrified, by all accounts, and ordered Martini to destroy the document at once. He added that the most stringent precautions were to be taken to prevent any leakage about the German trials. All experiments with the strips were to cease forthwith – even those directed at evolving an antidote to *Düppel* – i.e. 'Window'. As General Martini later commented: 'It was thus extremely difficult to work out countermeasures because we dared not experiment with the little beasts for fear of their being discovered. Had the wind blown when we dropped the metal strips, people would have picked them up, talked about them, and our secret would have been betrayed.' At all costs the British had to be prevented from learning the possibilities of this simple tactic.[4]

Sir Charles Portal had called for the postponement of

4. See page 141.

the introduction of 'Window' pending trials of its effect on British radar. Lord Cherwell had backed Portal, and urged that an antidote should be developed first. But there was a powerful lobby arguing for the immediate introduction of the 'Window' tactic into bombing operations: in this RAF Bomber Command was supported by Air Vice-Marshal Sir Norman Bottomley, the Deputy Chief of Air Staff. Bottomley felt that jamming of *GEE* could be expected to commence within the next three months, now that *GEE* sets had been lost over Germany; so the RAF should exploit all its tactics while it could. He felt it unlikely that the German air force could retaliate before that winter, and by then the much sought-after antidote to 'Window' might exist. He was supported in turn by Dr Robert Cockburn, head of TRE's jamming section at Malvern, who rightly believed that the *Würzburg* was the king-pin of the German air-defences, and said so forcibly in a paper circulated at this time:

(*a*) every 109 Squadron investigation aircraft on any flight near or over enemy territory has been continuously held by two or three 53-centimetre [570 megacycle, the *Würzburg* frequency] beams. It is an obvious inference that all aircraft operating over enemy territory are followed by 53-centimetre beams.

(*b*) from the serial-numbers of captured equipment and from other data, it is almost certain that this equipment exists in hundreds.

(*c*) the known characteristics of the 53-centimetre equipment are such as to allow operation as an SLC [searchlight control], GL [gun-laying] or GCI [ground-controlled interception radar].

(*d*) positive proof of its use as an SLC and a GLC has been obtained by 109 Squadron aircraft. It is perhaps relevant to mention that five aircraft have been lost during this investigation.

Cockburn suspected that the opposition to the immediate introduction of 'Window' sprang from an impression that its use would have a negligible effect on losses: 'If this is the current official opinion, we must take immediate action to alter it.' But the current official opinion remained unchanged, and perhaps it was for the best.

While the great controversy was raging over the use of 'Window' a similar controversy was arising over one of the two new radio devices being developed to improve the RAF's bombing accuracy, under the code-names *H2S* and 'Oboe'. The first of these bombing aids, *H2S*, depended upon the high-power 'magnetron' valve built originally by Professor J. T. Randall and Mr A. H. Boot in February 1940. This remarkable new valve generated no less than 500 watts of power on the high frequency of 3,000 megacycles – a scientific break-through which meant that radar sets could be developed capable of plotting targets with far greater precision than had been previously possible.[5] By July 1940 the power of this secret magnetron valve had been increased to ten *thousand* watts, and in the following month an experimental 'centimetric' radar using the valve had detected an aircraft at a range of six miles. In March 1941, a magnetron powered radar set was carried aloft for the first time, the prototype of a new radar for night-fighter aircraft.

Bomber Command's interest was aroused by reports that during the flight trials with this centimetric radar, operators had noticed that echoes from the ground were returned most strongly by built-up areas, and most weakly by the flat countryside or sea. As Bomber Command's targets were 'built-up areas' for the greater part, the question was raised whether the new night-fighter radar could be modified to 'look' at the ground, and installed in bombers. This was the birth of *H2S*. The rotating aerial system was modified to tilt the beam downwards, and the echoes were displayed on a cathode-ray tube in such a way that the sweeping radar trace sketched out a passable representation of the

5. The frequency of 3,000 megacycles per second corresponded to a wavelength of ten centimetres. The radar sets using the magnetron accordingly became classed as 'centimetric' radar.

surrounding countryside, just like a map. The modified radar underwent its first trial towards the end of 1941: soon after the take-off, Dr O'Kane, the radar set's operator, observed a large bright spot on his screen – the city of Southampton. The absence of echo signals from the sea threw the coastline into sharp relief.

It was clear that this was a device which would go a long way towards solving Bomber Command's main problems. The new set was independent of beams and beacons, so its range was in effect limited only by the range of the aircraft itself. It was this very success of the ultra-secret magnetron valve in the prototype *H2S* that brought up the old dilemma, which will by now have become familiar to students of the history of radio countermeasures: what would happen if the Germans were to capture one? If they were to copy the design and incorporate it in their own centimetric radar, and install this in their night-fighters as the RAF had installed it in theirs, their defences might enjoy increasing success against the British night attacks.

Ideally some sort of demolition device should be fitted to the secret valve to prevent its falling into enemy hands; the only alternative was to build *H2S* sets using not the magnetron but the less satisfactory 'klystron' valve, which worked on the same frequency but on much lower power. The magnetron was not so easily demolished: its main body was a block of copper, with a labyrinth of cavities machined out of it. The primary difference between the previous, low-powered versions and the new secret one was the manner in which these cavities were inter-connected, and that could not be easily camouflaged. Explosives experts at Farnborough devoted considerable thought to the fitting of a suitable destructor to the block and its system of connections, so as to render them unrecognizable. It had been worked out that two ounces of explosive would be necessary, but during a trial using a captured Junkers 88 as a 'guinea pig' aircraft, the destructor blew a ten-foot hole out of the fuselage, and it was still possible even then to tell how the magnetron worked by studying its fragments. Another idea tried was that of explosively ejecting the magnetron, and blowing it up while it was in mid-air. This too failed: this time it was the recoil force, which threatened to break the very aircraft up. The use of powerful acids was considered, but dismissed as impracticable. Finally, the idea of using the klystron to power *H2S* was also abandoned during July 1942, so it would have to be the magnetron or nothing. This time the offensive spirit won: Bomber Command's difficulties had to be overcome somehow, and this was the only way. The magnetron valve was readied for service in *H2S* at high priority, and the production of the sets themselves began. By the end of the summer, all idea of installing an effective demolition charge was abandoned. Sooner or later the Germans were bound to capture one intact.

While the effective operating range of *H2S* was virtually unlimited, a second device had also been completed during the summer of 1942, but its range was strictly limited. To compensate for this, its accuracy was very high indeed and was such that aircraft which were equipped with it could mark or bomb through cloud and overcast to the highest degree of precision. This system, code-named 'Oboe', was also developed by the Telecommunications Research Establishment. 'Oboe' made use of the fact that a radar set can measure the distance of an object to within fine limits, and that this accuracy does not diminish with range. Two transmitters were used, one at Dover and the other near Cromer. They employed radar-type transmitters, which sent out a stream of pulses to the aircraft; the pulses triggered a special airborne transmitter, which replied with transmitted pulses of its own. By measuring the distance between the first transmitter and the aircraft in the normal radar way, it was possible to direct the pilot by means of radioed instructions to fly along a circular path centring on the transmitter at Dover. The second transmitter, at Cromer, also used the radar principle and a repeater-transmitter in the aircraft to follow its exact position along the circular path controlled by the Dover transmitter. When the Cromer station observed the aircraft to be at the correct computed bomb-release point, it radioed the 'bomb release' signal. The short circular path actually flown by the aircraft was of such a large radius that it was virtually straight; for targets in the Ruhr, this path lay almost due North and South across the targets.

The 'Oboe' system promised extremely good bombing accuracy anywhere within its maximum range, determined by the curvature of the earth; the maximum range, 270 miles if the aircraft flew at 28,000 feet, took in most of the Ruhr with ease. The system's only limitation was that each pair of ground transmitters could only control one aircraft at a time, so obviously the device was best employed by Bomber Command's Pathfinder Force, which had been set up in August to emulate the methods of the German *Kampf-gruppe 100*; indeed, 'Oboe' had certain similarities to the latter squadron's *X-Gerät* system. No. 109 Squadron's Mosquitoes were the first aircraft to be fitted for the use of 'Oboe'.

In the meantime, the great controversy over the proposed use of tinfoil strips – 'Window' – to jam the German *Würz-burg* radar system had continued unabated between Bomber Command and Sir Sholto Douglas. By the middle of July 1942, Douglas had the results of the trials using 'Window' against each of the British airborne and ground radar sets: it had jammed every device against which it had been employed. Dr Derek Jackson, of the Telecommunications Research Establishment, and the other experienced radar operators had all expressed the view that no amount of practice would serve to improve the position.

The Chief of Air Staff wrote to Douglas about the serious dilemma created by the trials: while on the one hand it was possible to assume that the 'Window' idea had not occurred to the Germans, on the other it was equally possible that they *had* had the idea but shrank from using it while the RAF's bomber strength was the greater in the west. The danger was that the very act of training Fighter Command to operate in face of 'Window' jamming might compromise the secret with the enemy. On 21st July, Portal called a meeting to air the whole question. Air Commodore Tait, Sir Robert Watson Watt, Air Vice-Marshal F. F. Inglis (Assistant Chief of Air Staff, Intelligence), Lord Cherwell, Dr Jones and Sir Sholto Douglas were invited to attend. Jones submitted to the conference that the idea of 'Window' was so simple that it was unlikely that the Germans had not thought of it. If they had readied it for operational use, then the German air force was withholding it until they resumed large-scale attacks on England. But he was inclined, on the scanty evidence then available, to hold that they had not readied 'Window' for operational use. Lord Cherwell disagreed, expressing the view that the Germans had not in fact thought of 'Window', and it was important to conceal the fact that Fighter Command was conducting trials with the tinfoil strips. As one could never kill rumours completely, he thought the most effective step would be to circulate a counter-rumour that the RAF had tried the strips, but found them to be quite ineffective. Inglis was instructed to take the necessary action.

The Chief of Air Staff summarized their decisions in a paper circulated to the Chiefs of Staff committee on 30th July. He told them that the trials had produced sufficient evidence to show that it would be inadvisable to use 'Window' until the British radar system had developed a degree of immunity to it – always assuming that the Germans did not use it first. He had asked the C-in-C of Fighter Command to press the development of night-interception tactics which did not rely on radar, and he had asked Air Commodore Tait to see what could be done to provide a more efficient sound-locator set to back the radar network. For the present, he had ordered that the 'Window' tactic should not be used over Germany.

Three months later, vague news reached the British Intelligence services of German work on a tactic identical to 'Window'. Dr Jones received a report from an agent who had struck up a conversation with a German air force woman auxiliary on a train. The woman had told the agent that she worked at a night-fighter control station, and she made an obvious reference to radar – a detector able to 'see' metal in an aircraft even in darkness. With a little prompting, the woman had related how on one occasion a British bomber flying over the Rhineland had deceived the 'detectors' by throwing out 'aluminium dust'.

Jones did not think that the agent had made the story up – such men had not proven good inventors in the past. Nor did it seem likely that the German Intelligence authorities could possibly gain anything by gratuitously presenting the British with the 'Window' principle. The only reasonable assumption was that the woman concerned had heard a garbled account of the German air force's own trials. Jones called on Lord Cherwell to discuss the Intelligence report on the evening of 3rd November: Cherwell's original objections to the introduction of 'Window' by Bomber Command had been based on the belief that the Germans had not thought of it, and now it seemed that they had. Jones had no wish to see his old professor appear in a false position at the top-level conference both would be attending on the following day. Lord Cherwell would have none of it, and perhaps not unreasonably said it would be silly to change Air Staff policy on the basis of something someone had said on a German train. As he was leaving Cherwell's rooms, the latter said: 'If you go into the meeting and try to get "Window" used, you'll find me and Tizard united against you.' Jones, knowing the famous barrier of mutual distrust existing between Cherwell and Sir Henry Tizard, retorted: 'Well, if I've achieved that, by God I've achieved something!'

Air Vice-Marshal Saundby represented Bomber Command at Portal's conference next day. At Cherwell's request, Derek Jackson – by now a Wing-Commander – had worked out in detail how 'Window' could best be used to defeat the German radar; he had also considered what steps were necessary to make the British network less vulnerable to it. Sir Charles Portal was sufficiently impressed with Jackson's plan to put him in charge of a new body, the 'Window Panel', to assume responsibility for the development and production of the material and the RAF's own countermeasures to the tactic.

Jackson reported that the latest British night-fighter radar, the AI Mark IX, had an automatic target-following device which should make it easier to distinguish between aircraft and 'Window' echoes; this radar should be ready for service by mid-1943, and the prototype was due to fly soon. There was also an American night-fighter set, the *SCR720*, which employed a system of scanning and display which might enable it to work in the face of 'Window'; one set had been ordered, but had not yet arrived. On the ground there was a new fighter-control radar, Type II, operating on a frequency of 500 megacycles. Although the set did not employ any special anti-jamming circuitry it could still be used in a 'Window' campaign, as it had been kept rigidly in reserve in order that the Germans should not learn of its existence. Six had been delivered, and forty more of an improved model were being produced under a crash programme. The investigation of improved means of sound location on the other hand had proven fruitless.

All now seemed set for the early use of 'Window' to jam the German *Würzburg* system. But there was a surprise in store, for Air Vice-Marshal Saundby now expressed himself reluctant to press for the use of 'Window'. He explained that there would only ever be a limited number of tricks that the force could use against the enemy. Once these had been exhausted, there was nothing; and no trick however clever was likely to keep the Germans 'on the hop' for very long. There was a lot to be said for playing each new innovation until it had been played right out, and only then moving on to the next one. This was the time when 'Mandrel' and 'Tinsel' jamming was about to commence. 'Window' should wait until the Germans had mastered these.

This was an entirely unexpected argument. It pulled the carpet from beneath R. V. Jones' feet, and left the floor to those who opposed the tactic's introduction. Once again the decision was made to withhold the use of 'Window'; and once again Fighter Command was asked to advance its experiments with countermeasures to the German use of 'Window' jamming, Sir Charles Portal said that he would review the matter – in six months' time.

*

The year ended in a tragedy for the Telecommunications Research Establishment. Early in December 1942, the prototype AI Mark IX radar took the air in a Beaufighter aircraft. Wing-Commander Jackson was the operator. During this first flight, Jackson found that the new radar tended to lock itself on the 'Window' clouds instead of the proper target, the 'enemy' aircraft. Dr Downing, the physicist who had been responsible for much of the Mark IX's design and construction, modified it successfully to overcome the fault.

On 23rd December, the prototype was again ready for air testing. This time Downing asked to fly with Jackson, so that he could see how successful his modifications had been. As there was nobody available to launch the 'Window' from the target aircraft, Jackson decided at the last moment to let Downing operate the new radar, while he himself dropped the metal foil from the observer's position in the target aircraft. As the two Beaufighters neared the appointed area, Wing-Commander Jackson saw a Spitfire fighter aircraft swinging round on to their tails in a most aggressive manner. He warned his pilot, who threw the heavy Beaufighter into a steep diving turn, but almost at once it shuddered under the impact of cannon fire from the Spitfire. Jackson tried to speak to his pilot, but could get no reply; in fact the cannon shells had disrupted the intercom between them. Each thought the other had been killed. For Jackson the only way out now was to bale out, but the diving turn was so fierce that the 'G' forces prevented him from reaching his parachute. For ten terrifying seconds he waited for the end. As the Beaufighter levelled out a few hundred feet above the water, he saw the fault in the intercom. He plugged in again just in time to hear the last snatch of a message his pilot was broadcasting over the radio-telephone: '... and he has killed my observer!'

Their aircraft was so badly damaged that they had to return to Coltishall at once. Any doubts as to what had become of the Spitfire after it had attacked them were dispelled as they turned back towards the land: there was an aircraft burning in the sea – but it was not the Spitfire. It was Dr

Downing's Beaufighter, bearing the precious prototype of the new radar Mark IX.[6] Downing himself was dead.

As you will see, the Mark IX seems to be the answer [Jackson wrote to Lord Cherwell five days later]. Unfortunately it has now been lost: I was doing a final test just before Christmas, and the two aircraft we were using were both attacked by a Spitfire. He shot down the Mark IX and damaged, but did not shoot down, the Beau that I was in.

I do not know to what extent this will hold up the production of Mark IX, but it is a most tragic affair. One of the TRE men was in the aircraft shot down, and he was the expert on Mark IX.

The real tragedy was that Downing's sacrifice had been in vain. When the first example of the new American radar *SCR720* arrived in England that same month, it was seen that it was superior to the Mark IX. When the 'Window' was released, the tinfoil strips normally took some ten seconds to blossom into a fully reflective cloud, during which time the bomber would have moved about half a mile. The new American radar accordingly had two screens: with one of these the operator would pick out the real target at the head of the stream of 'Window' clouds; at the turn of a knob, he could move the 'blown up' area covered by the second screen until it coincided with the target aircraft, and follow it along with little interference from 'Window'. The *SCR720* was ordered in quantity. Wing-Commander Jackson himself flew the initial trial flights, and on 4th May 1943 it was to make its first operational flight on night patrol, with Jackson again acting as observer in the Mosquito in which it was fitted. At the end of that month, the first of a batch of 2,900 were due to arrive in Britain from America's Western Electric Company. The new radar was designated AI Mark X, and Downing's Mark IX was

6. An investigation later revealed that the Spitfire's pilot was a Canadian on his first operational flight.

dropped. The last sound objection to the use of 'Window' by RAF Bomber Command had disappeared.

The RAF's concerted jamming of the German night-fighter defensive system began at the end of the first week in December 1942, when a force of bombers was sent to raid Mannheim. Both the 200-mile long 'Mandrel' screen deployed to jam the *Freya* early-warning system and the crude 'Tinsel' jamming of the night-fighters' radio-telephone communications were employed. This was an unpleasant innovation for the Germans. Their night-fighter force's diarist noted: 'Heavy jamming of *Freya*. It was nearly impossible to control the night-fighters.' Several interceptions were hampered to such an extent that the bombers had passed through the line and out of range of the Giant *Würzburg* radar sets before the German fighters could make contact. That night Bomber Command lost nine aircraft, only 3·3 per cent of the attacking force.

As the initial shock of 'Mandrel' passed, the German radar operators got used to the jamming. As professional men will, they soon devised means of avoiding the worst of it. The simplest way was to 'de-tune' the radar away from the main jamming frequency. It was only a temporary palliative, but it did enable the radar operators to see through some of the jamming.[7]

In a more permanent attempt to nullify the effects of 'Mandrel' jamming, the German *Freya*, *Mammut* and *Wassermann* sets were modified to operate on frequencies spread over a much wider range than before. Originally the range had been from 118 to 128 megacycles; now it was stretched from 107 to 158 megacycles. This did nothing to reduce the efficiency of the radar, but a single 'Mandrel' jammer could

7. This technique – 'de-tuning' – has a counterpart in everyday life. Sometimes two domestic radio programmes may appear so close together that they blot one another out. By re-tuning the receiver slightly and sacrificing some of the volume from the required station it is often possible to bring about a considerable reduction in the interfering signals.

transmit noise only over a ten-megacycle wide band of the radar frequency spectrum. For a given number of jammers, the power radiated on any one frequency was accordingly considerably reduced.

A further problem developed when it was realized by bomber crews that German night-fighters might home on to the 'Mandrel' jammers' aircraft. To prevent this, the 'Mandrels' were further modified so that they radiated their noise for two minutes, and then went silent for two minutes, radiated for two more minutes and then went silent again, and so on. But this reduced the overall strength of the jamming still further, since on average only half the 'Mandrels' were transmitting at any one instant. By the end of the first three months of the new year the Germans were over the worst, and the RAF's losses started to rise again. The German early-warning radars, operating on the new frequencies, were able to see through the old 'Mandrel' screen, and since the small Defiant aircraft carrying the 'Mandrels' lacked the room for additional jamming equipment the screen's operation was discontinued for a while.

The noise jamming of the night-fighters' R/T frequencies[8] was a greater and more lasting success. At first it had caused much confusion in the German ground-control system, and RAF wireless operators who understood German reported lengthy repetitions of orders and signs of growing irritation among the German aircrew. The immediate enemy reaction was to increase the power of the ground-control transmitters to make it easier for the night-fighter crews to hear the orders above the jamming. Then the RAF monitoring service heard the Germans broadcasting instructions to their night-fighters on frequencies between 38 and 42 megacycles, a range previously used only by the enemy's day-fighters. It was obvious that the Germans were avoiding the noise jamming by fitting additional radio equipment to all their night-fighters. But the older R/T sets were still in use, and their jamming continued until the war had ended.

There was an important lesson to be learned from the 'Mandrel' and 'Tinsel' episodes, a lesson which must be

8. Code-named 'Tinsel'.

borne in mind when the value of an electronic jamming offensive is considered: electronic warfare is essentially a fast-moving campaign, in which the victories are relative, not absolute. Enemy measures, both defensive and offensive, might be hampered or delayed, but they could never be absolutely confounded. Given time, a resolute enemy would always be able to introduce new equipment immune to the form of jamming in use. These two simple jamming devices had still achieved a relative success, however: at small cost they had forced the Germans to modify most of their early-warning radar sets to cover new frequencies, and to fit new radio sets to their night-fighters. Dr Cockburn's dictum that a 'shilling's worth of radio-countermeasures will mess up a pound's worth of radio equipment' had been proven correct. Several months of confusion and anxiety had been inflicted on the enemy, during which about one hundred RAF bombers and their crews had been saved from the end that they would otherwise have met.

6

The Rude Awakening

I expected the British and Americans to be advanced, but frankly I never thought that they would get so far ahead. I did hope that even if we were behind we could at least be in the same race.

– REICHSMARSCHALL GÖRING in May 1943

The deliberate jamming of the German *Freya* early-warning radar system should have sufficed to bring the German high-frequency electronics experts to their senses; but when they first realized that the systematic jamming of *Freya* had begun, in December 1942, their reaction was one of gratitude that the British had not found any means of dealing with the *Würzburg* radar system too. It is hard to understand why they should have nurtured this belief. They knew that the *Würzburg* had been the object of a specific – and successful – combined operation by the British earlier that year, and they knew that they themselves had devised the simplest method to make the *Würzburg* radar unworkable – the release of tinfoil strips in the radar's beams.

When the possibility that the *Würzburg* system might be jammed by the enemy came under anxious discussion at a meeting of the German night-fighter controllers, General Kammhuber, C-in-C of the night-fighter command, 'referred to a folder in his possession and informed the gathering that according to a report which he had received from the Experimental Station at Werneuchen, such jamming was impossible'. Kammhuber subsequently explained in private that he knew full well that this was not so, but he had had to obey Göring's earlier decree forbidding any further discussion or research into the German equivalent of the 'Window' technique, code-named *Düppel* by them.

One basic factor contributing to the weakness of the German radar defences at this time was the continuing rivalry between Field-Marshal Erhard Milch, Director-General of Air Equipment and one of the more colourful characters of the German Air Ministry, and General Martini, who was Chief of Air Signals; the latter was a scholarly and cautious officer, with little understanding of the development of radar, and Milch had no time for him. Martini was also unequal to the intrigues which were woven round him. When Göring asked some months later, who was really responsible for the German radar industry, both Milch and Martini said it was no concern of theirs.[1] Göring decided it should be Milch's responsibility, but Martini continued to interest himself in it as before.[2]

To confuse the issue still further, Staatsrat Dr Hans Plendl, the expert who had devised the early bombing-beam systems used against Britain, was appointed Göring's Plenipotentiary for High Frequency Research on 14th November 1942, and this clearly meant he would be overlord over the radar programme. The picture which confronted him as he took up his new commission was a depressing one: Germany had only one-tenth of her enemy's radar research capacity, and this was scattered around a hundred minor

1. The curious relationship between Milch and his Reichsmarschall is well illustrated by an incident in the summer of 1941, when Milch had suggested that the German air force should build up an integrated fighter-control system like that used by the RAF. During a later discussion on air defence, he had reminded Göring of the memorandum in which he had made the suggestion. The latter had replied: 'Don't flatter yourself that I read the rubbish you send me!'

2. Field-Marshal Milch announced at a meeting in Berlin on 6th July 1943 that the Air Ministry was taking over all radar and electronics research, on Göring's orders: 'General Martini alleges that this has always been the case. As far as I am aware, we have somewhere a written order that this was Martini's province. Martini disputes that. In any case, it comes to us and we carry on with it. Development and research in the whole field of electronics are now our concern. The order has been given.'

institutes and organizations, each working independently of the other. Plendl's first action was to secure a Hitler Order recalling 1,500 scientists from the Front to staff a small number of specially established research centres. But all this was happening too late, as the enemy now had the initiative.

The Germans still believed that they were ahead of the British and Americans in radar technology. In particular, they were reluctant to retaliate for the heavy jamming which the British had begun. Field-Marshal Milch inquired at an Air Ministry conference called on 5th January 1943, whether the Germans might not develop similar methods of jamming the British radar. Engineer-Colonel Schwenke, the Intelligence expert, told him: 'That is basically possible.' Milch said that the matter should be put in hand at once, but Colonel von Lossberg, Milch's night-fighter expert, interrupted with an ominous objection: 'May I say here that this has been discussed with General Martini, and he has urged that all attempts to jam enemy radar traffic should be ceased for the present, as there is a perfectly simple means of jamming our whole radar system – a means to which we have no countermeasure.'

General Galland said that if every RAF aircraft carried a jammer, all he would have to do would be to fit his fighters with homing receivers and home in on the jamming transmissions. The trouble was that single aircraft laden with jamming transmitters disrupted all the radar system, while ordinary bombers went on to do their 'dirty work'. The fact that neither the British nor the Americans had started jamming *Würzburg* struck Colonel Schwenke as most unusual, and he said so. 'That is significant inasmuch as there are far fewer *Freya*-sets than *Würzburg*-sets in existence. The British must know the relationship of our *Freya*'s to our *Würzburg*'s.' Field-Marshal Milch summed up that they had to reach some kind of decision about jamming: 'Otherwise we will turn out *Würzburg*'s like mad and no *Freya*'s.' And he added: 'How will our fighter-control system operate in the coming year and the one after that, if we don't put our house in order?'

Two days later, there was a further tangible proof that

the Germans were losing their grip. It will be recalled that during the first year of the radio war, British Intelligence had discovered the German bombing-beams before they were first used in regular raids. Now that the RAF was beginning to employ its own long-distance radio bombing aids, the first that the Germans learned was when the raids began. On 7th January there was a meeting between disturbed air defence leaders in the Ruhr and directors of the Krupp works in Essen. It was quite clear to the people in Essen that ever since two days before Christmas, individual Mosquito bombers had been flying high across the town, almost due north and south, and dropping bombs with the highest precision regardless of the weather conditions. Was the enemy using some kind of infra-red homing device? Or a radio beacon, perhaps planted by agents in the town? Whatever the system, the Krupp workers were reported to be growing uneasy' about these sudden unannounced bombing attacks, as the sirens were not sounded for single aircraft.

Within the next twelve days, the suspicions seemed confirmed that the RAF had developed a highly accurate method of radio control for bombers, as all these high-precision attacks by lone aircraft had taken place within a radius of about 250 miles of the English coast. German electronics experts confirmed that if the Mosquitoes flew at 30,000 feet, radio beams could reach them from England, but there were still many who refused to believe that this could be happening to Germany.

The RAF had begun to use 'Oboe' operationally during the night of 20th December, when six aircraft used it to attack a power station in Holland. During the week that had followed, 'Oboe' Mosquitoes had gone out each night, in ones and twos, to demonstrate its remarkable accuracy against pin-point targets, and on Christmas Eve one of the aircraft had carried out a lone attack on the Ruhrort steelworks.[3] But 'Oboe' raids were not the only developments to

3. After the Duisburg–Ruhrort raid, a German news broadcast announced: 'Some British aircraft broke the peace of Christmas night and attacked western Germany. Among other objects, several

put the Germans ill at ease in the first three weeks of 1943.

By the middle of January, ten Halifax bombers of No. 35 Squadron and thirteen Stirlings of No. 7 Squadron had been fitted with the *H2S* radar equipment: their radar operators could see on the screens the brilliant areas of the towns, and the darker patches showing the lakes and river estuaries. Both squadrons were units of the RAF's own Pathfinder Force, formed in August 1942 to emulate the tactics and the success of the German *Kampfgruppe 100* over England. As a result of a herculean effort by the Telecommunications Research Establishment and the ground crews at the airfields, the new radar was ready for action by the end of the month. It was first used operationally on 30th January during a night attack on Hamburg.

Now the Germans were at the dawn of a very rude awakening indeed. During the second of the RAF's attacks using *H2S*, the Cologne operation on 2nd February, a German night-fighter shot down one of No. 7 Squadron's Stirlings near Rotterdam in Holland. When German technicians began a routine investigation of the wreckage next day, they found something that was completely new to them. The secret of *H2S* was out.

Schwenke reported the discovery to a top-level Berlin conference of technicians and staff officers nine days later. By now a more than cursory examination of the equipment had shown that the British had found ways of operating on a far higher frequency than the Germans had believed possible.

'I have to report,' announced Schwenke, 'that a new device has been found in a Stirling bomber shot down near Rotterdam. It is a centimetric radar device mounted beneath the rear of the fuselage. We have not yet established what exactly it is, but the device is exceedingly costly. This is the first appearance of a decimetric apparatus, for the British

---

graves in a remote cemetery were destroyed by the bombs.' The broadcast did not add that the cemetery in question was right next to the Ruhrort steelworks.

have hitherto given no hint of much progress in this field ...
Two units are missing from the equipment, out of a total of
six or eight.' Field-Marshal Milch asked whether that meant
that the units had been lost, and Colonel Schwenke replied:
'We salvaged the equipment from an eighty per cent wrecked aircraft. The two missing units were not with the rest
of the equipment, but were probably mounted up front with
the pilot. For a time Lorenz and other experts believed it
was a night-fighter warning set, but it is too complicated for
that. It is suspected therefore that it is a night-fighter search-
or warning-device, and simultaneously a navigation and
target-finding device.'

Schwenke added that there was a possibility that the
equipment might be used for remote-controlled take-off,
although no connection with the actual controls had been
found. Apparently the night-fighter crew which had shot the
bomber down had noticed nothing unusual. Two of the
Stirling's crew had survived the crash but 'both have obsti-
nately and consistently refused to make any kind of statement; this strongly suggests that it is something special.'
There was adequate proof also that the set was entering
mass-production.

The captured *H2S* units – by now code-named '*Rotterdam*' by the Germans – were turned over to the Telefunken
company in Berlin, who had designed and built the *Würzburg* radar system to which the new units appeared to have
some similarities. On 22nd February, General Martini set
up a special '*Rotterdam* Commission'[4] to devise and develop
the necessary countermeasures, and the first session, with
Staatsrat Plendl and experts from industry and the services
present, took place in the Telefunken works that day. The
awful realization of the great advances made by the British in
centimetric radar technique hung heavily over the meeting.
Plendl later reported that in Germany even the most basic
research in these techniques was only in very early stages,
'although German researchers had repeatedly drawn attention to the importance of such work'. It was small wonder
that it took many weeks to find out how the new radar

4. Arbeitsgemeinschaft '*Rotterdam*'.

worked, and months to discover what kind of picture it provided.

At the first session of the '*Rotterdam* Commission', the Telefunken works undertook to manufacture six radar sets on the lines of the captured *H2S* as prototypes for a mass-production model; at the same time, basic plans were discussed for two *H2S* detection devices, one of which – *Naxos* – was to be a simple detector of centimetric waves, and the other a direction finding receiver code-named *Korfu*.

On 1st March, the precious *H2S* units salvaged from the Rotterdam crash were completely destroyed during a heavy attack on Berlin, in which the Telefunken works was badly damaged. But ironically one of No. 35 Squadron's Halifaxes was brought down over Holland during the same night, and a new *H2S* set fell into the Germans' hands. Again there was no cathode-ray display unit among the units which were salvaged, but at a further '*Rotterdam* Commission' meeting three weeks later, attended now by nearly thirty experts, an Air Ministry official reported that an RAF prisoner had confirmed that *H2S* was a navigation aid used by the Pathfinder Force for releasing its marker flares.

On 23rd March, Colonel Schwenke told Milch about the new find: 'The sets which have fallen into our hands have so far lacked their display unit, that is the unit containing the Braun tube [i.e. cathode-ray tube]. But the interrogation of the prisoners has revealed that the device is certainly used to find targets, inasmuch as it scans the territory over which it flies on a Braun tube. The features of the countryside are shown bright or dark according to the intensity of the echoes: buildings and forests are bright, and all smooth spaces like water, fields, etc., are shown as darker. Further interrogations of prisoners have revealed that the set is used not as a navigation aid but for identifying countryside reached by other means of navigation. This obviously means that the camouflage of decoy sites is seen right through by such equipment.' Schwenke proposed that special anti-*H2S* camouflage be developed, using simple wire-netting to send back stronger echoes, perhaps. However, there was still a lot to be learned about exactly what the British saw on their *H2S* screens.

In the meantime, the newly captured *H2S* set was being reassembled by Telefunken engineers in one of Berlin's huge concrete flak-towers, where it was safe from the heaviest bombing attack. There its secrets were laid bare, and the same excitement as had gripped the Farnborough engineers on first making their captured *Würzburg* radar work was sensed by the Germans, as they saw on a flickering screen improvised from German equipment, a representation of the city around the huge Berlin flak tower. The truth about *H2S* was coming as a great shock to the Germans: the magnetron which alone gave usable power at a frequency they had previously considered useless, the plan position indicator form of presentation and the special circuitry for generating extremely narrow pulses of energy – all made a profound impression. For once even Reichsmarschall Göring was subdued. After reading an initial report on *Rotterdam* in May, he remarked: 'We must frankly admit that in this sphere the British and Americans are far ahead of us. I expected them to be advanced, but frankly I never thought that they would get so far ahead. I did hope that even if we were behind we could at least be in the same race!'

For the Germans, the mystery of the precision attacks by Mosquitoes on Ruhr targets still remained: now that *H2S* had been found it seemed too bulky to install in the Mosquito bomber, so the British must have something else. On 5th March 1943, the first heavy Bomber Command raid to rely on 'Oboe' Pathfinder-marking took place, and again the target was the Krupp works in Essen. In the past, the works had been a difficult target to find: the thick industrial haze seemed always to hang over the city, and this night was no exception. But at precisely 9 p.m. the first of the 'Oboe' Mosquitoes marking for the attack released its salvo of red target-marker bombs, unhampered by the haze. Three minutes later a second Mosquito followed suit, followed seven minutes later by a third and so on throughout the thirty-three minutes of the attack's marking phase. Twenty-two Pathfinder Force heavy bombers backed up the marking with green target indicators, but the main force crews were told that as the method of placing the red markers was 'a new and very accurate one' they were to attack these where possible, and with the greatest possible precision. At no stage was visual identification of the Krupp works called for.

The result was a resounding success for Bomber Command, and the Krupp works was very heavily damaged. Shortly after the second such attack on Essen, General Martini was summoned to Hitler's headquarters. On arrival he found himself the centre of a discussion on the RAF's latest bombing attacks. The Führer said he had read that the Krupp works had been hit during a night attack and through continuous cloud cover; it had been suggested that a new British radio navigational aid was responsible. Was this possible? Martini replied that it was.

Göring, sensing trouble in store, interjected: 'Yes, my Führer, but we also have such systems.' Hitler asked for details, and Martini outlined the *Y-Gerät* system used in 1941. Hitler said: 'I want to know whether, if you were to attack Munich's main railway station from Leipzig with your system, you could hit it?' Without committing himself too deeply, Martini replied: 'I should estimate that Munich is about four hundred kilometres from Leipzig. If that is so, and if the station is 1,000 metres long by 300 wide, then I believe that some of the bombs would hit the target.' 'I hope this is correct,' commented Hitler, 'I don't trust high-frequency gadgets. I once went on a flight in southern Germany and ended up in northern Germany by mistake, all because of your high-frequency gadgets.' He ordered a demonstration to be carried out with the *Y-Gerät* over Germany, under operational conditions, to show whether these things could really be done.

In due course the demonstration was carried out, under what Staatsrat Plendl later described as 'warlike conditions'. It resulted in the release of 50 per cent of all the bombs within a radius of 900 yards at a range of 225 miles – a fair result for the range, but not up to 'Oboe'; the target area was an uninhabited site near Bayreuth. Martini, careful not to make the trial too realistic, omitted the jamming which had been the system's downfall during operations over England.

The demonstration seemed strong circumstantial proof that the British were using a radio-beam device in their attacks on the Ruhr, but no such equipment had been recovered from any aircraft shot down. This was hardly surprising, since the Mosquitoes were almost immune to the German defences by virtue of their high performance.[5] Moreover 'Oboe', like *GEE*, had a very effective self-destruction device fitted to it. 'Oboe's' unusual signals had been picked up by the German monitoring stations along the northern coast of France as early as the autumn of 1942 – it was noted that they generally occurred at night and seemed to be in some way associated with German E-boat activity in the Channel – but these were naval monitoring stations, and the reader need not be surprised to learn that the signals were not reported to the German air force. When at the end of May 1943, Wuppertal-Barmen was the target for a particularly precise RAF attack in which three thousand civilians lost their lives, the anti-aircraft defences clearly made out the arrival at very high-altitude of a lone Mosquito aircraft three minutes before the main bomber force arrived. The defenders watched the release of the Mosquito's markers, but still nobody had intercepted any signals associated with any kind of blind-bombing system. Two more months were to pass before a German Post Office expert, Dr Sholtz, was able to correlate the signals heard by the Navy with actual dropping of the Mosquito's target markers during a large-scale attack on Cologne. And in the meantime, 'Oboe' was bringing destruction to the Ruhr on a very large scale indeed.

Why were the British still not jamming *Würzburg*? The matter had again been raised by uneasy Intelligence Officers

5. During the first six months of 1943 only two 'Oboe' Mosquitoes were lost; neither fell into German hands.

in Berlin at the end of the third week in March. A captured
British document instructing ground forces in the best
manner of disabling the *Würzburg* radar installation had
also contained details of the set captured during the Bruneval raid, so it was clear that the British knew both about this
and the *Freya* systems: 'It is a source of puzzlement for
us,' reported Colonel Schwenke to Milch, 'that the *Würzburg* radar is not being jammed.' What had caused the matter to be raised again was the discovery of two new devices
in the wreckage of British aircraft, showing that the RAF
was aware not only of the *Würzburg* radars on the ground,
but also of the *Lichtenstein* airborne radar installed in the
night-fighters. Colonel Schwenke reported:

On 2nd March, a directional aerial was found on the
rear of an aircraft shot down at Twente in Holland –
nothing else. The aerial gave immediate reason to believe
that it was part of an active night-fighter warning device.[6]
Then, on 12th March, we found a further aerial in another
Halifax near Münster, with part of its mounting still
attached. The tags on this mounting are labelled 'receiver'
and 'transmitter' . . . which proves that it is an active
warning system, which radiates radar beams at the enemy
aircraft and employs the returning energy to detect it.

The Germans had in fact found pieces of the newest
Bomber Command device, code-named 'Monica': this was
a rearward-looking radar-set, designed to give warning of
fighter aircraft in the 45-degree wide cone extending a
thousand yards behind the bomber. This and another device,
code-named 'Boozer', entered service in the spring of 1943;
so the Germans had certainly found out about it very
quickly.

'Monica' was designed to give warning of aircraft approaching from the rear by sounding a series of bleeps into
the crew's earphones, rapidly increasing in rate as the range

6. An *active* radar device is one which both transmits and receives signals. A *passive* device, on the other hand, only receives
signals.

closed. The device gained immediate unpopularity because of the constant false alarms it gave from friendly aircraft in the bomber stream. 'Monica' was one of two warning devices being fitted to RAF aircraft at this time. The second was 'Boozer'.

'Boozer' was a *passive* device, consisting of a radar-type receiver tuned to the frequencies of *Würzburg* and *Lichtenstein*, the German gun-laying and night-fighter radars. 'Boozer' could not give false alarms at all. When the aircraft was exposed to the invisible beams of a gun-laying radar set, 'Boozer' was designed to light up an orange light on the pilot's panel; if on the other hand a German night-fighter radar caught the bomber, a red warning light appeared in the bomber's cockpit. Even this system had its drawbacks, of course. The *Würzburg* warnings came too frequently to be of value, and there was no warning of night-fighters without radar; and when the red lamp went out this might only mean that the night-fighter had closed to visual range and had switched its radar off.

A 'Boozer' fell into the hands of the Germans on the same night as they acquired their second *H2S*. 'During the night of 1st March, a receiver fell out of a Lancaster shot down over Berlin; the receiver fell in someone's back yard,' explained Colonel Schwenke. 'They knew that it had had something to do with the aircraft, but it only reached us by a devious route on 12th March. It appears to be a broadband receiver taking in the frequencies of both *Lichtenstein* and *Würzburg*.' The virtually intact receiver had been examined by Telefunken experts, and RAF prisoners had been questioned; the latter had revealed that it operated coloured lamps to give warning when one or the other of the German radar sets was pointed at the aircraft. This afforded an interesting insight into British knowledge of German radar work, and into the lengths they would go to protect their own bombers, concluded Colonel Schwenke. 'It is still a mystery for us that the British are only jamming *Freya*, and not *Würzburg* and *Lichtenstein* as well.'

The mystery was soon to be dispelled. On 2nd April, Sir Charles Portal finally held the 'Window' conference he had promised six months earlier. Much had changed since the previous discussions: now an airborne radar, AI Mark X, was available which would not be completely dismayed by this form of jamming, and the new ground radar set, the Type II, was coming forward in useful numbers. Portal felt less strongly about the need to withhold the use of 'Window'. Concurrently with his trials with latest night-fighter radar sets, Wing-Commander Jackson had developed a 'Window' bundle with sufficient tinfoil to produce a 'heavy bomber' echo, while weighing less than two pounds. This meant that the bombers could now use the strips all along their route instead of just at the target, and the result would be far greater protection from the defences.

Sir Arthur Harris, C-in-C of Bomber Command, expressed the view:

> There is now a good possibility of saving one-third of our losses on German targets by using this countermeasure. The Command has nothing to lose and possibly much to gain by using it.

Harris carried the day, and Portal agreed to recommend to the Chiefs of Staff that 'Window' should be released for operations against the *Würzburg* radar system as from 1st May 1943.

In a report to the Chiefs of Staff committee at the end of April, Portal estimated that losses due to enemy action would fall by 35 per cent as a result of 'Window' jamming, and it would take the enemy six months to evolve new equipment to overcome it. On this basis, he calculated that had 'Window' been employed during the whole of 1942, 316 bombers and their crews would have been saved. This figure relied on the Germans not having been able to develop an effective countermeasure for one year following 'Window's' first employment – how unrealistic was this belief, was to be learned by Bomber Command to its cost during the winter of 1943. Portal was not alarmed by the prospect of the

Germans retaliating with the same tactic, as he expected fifty aircraft with the new AI Mark X radar to be in service by August, and eighteen Type II fighter-control radars by the end of the year. In any case, the great victory at Stalingrad had turned the tide in the east.

The small proportion of the bomber force that the Germans could now release for the western theatre of operations would not be able to sustain an effort of more than fifteen or twenty sorties a night against England. Portal concluded:

> The use of 'Window' will increase materially the effectiveness of our bombing offensive. The cost will be a possible increase in the effectiveness of the enemy's night bombing in this country and an increase in the difficulties of night air defence in overseas theatres.
>
> In the present strategic situation, however, the balance is overwhelmingly in our favour, and it is recommended that we should employ 'Window' as from 15th May 1943.

He asked the Chiefs of Staff for an early decision so that the necessary additional production could be put in hand. Now, however, there was a strategic argument against the early use of 'Window'. The Chiefs of Staff felt that if the Germans were to use 'Window' to support their air operations at the initial stage of the forthcoming Allied invasion of Sicily, planned for the beginning of July, the invasion might be jeopardized. The introduction of this method of jamming the German *Würzburg* system was accordingly to be delayed until after the beach-head had been consolidated; once this had been done, Bomber Command would be free to use 'Window'.

A few days later, the British Intelligence services were presented with the final link in their chain of information on the German night defences. On 9th May, a German night-fighter crew defected to Britain in a Junkers 88, and landed intact at Aberdeen complete with a working *Lich-*

tenstein fighter interception radar. So this was *'Emil-Emil'*, whose radiations had been monitored with such gallantry by the 1473 Flight 'ferret' aircraft six months before. The German aircraft was repainted in the markings of its new owners, and flown to Farnborough, escorted by a Spitfire; at Farnborough, it was employed in trial interceptions against British bombers. The radar was found to have a performance broadly comparable with the early British night-fighter radar equipment, although it did not compare with the latest. It could not operate effectively in the face of 'Window'.

The RAF was not to be the first service in the world to use metal strips to jam radar, as wheels had been turning ten thousand miles away as well. In Japan, Lieutenant-Commander Hajime Sudo, head of the Imperial Navy's radio countermeasures section, had developed *'Giman-shi'* – literally 'deceiving paper': this was made from the metal screening cut off electric cables, and backed with paper tape. Each strip was 75 centimetres long, half the wavelength of the American gunnery-control radar sets used in the Pacific theatre, and three millimetres wide. In May, as the battle for the Solomon Islands was nearing its close, the Japanese Navy started to use *'Giman-shi'* to protect its night bombers during attacks on Guadalcanal. While within enemy radar range, the Japanese bombers dropped bundles, each comprising twenty *'Giman-shi'* strips, at irregular intervals. The rate of discharge steadily increased until over the target one bundle was being dropped every five seconds. The tactics certainly deceived the American gunlaying radars, and the Japanese navy reported a reduction in their losses. But no report of these operations reached the British Intelligence services before the final decision was taken to use 'Window' over Germany. Dr R. V. Jones later said: 'It would indeed be interesting to know when the Japanese use was first reported back to Washington, and what happened to the information from there on, because it certainly did not come back to me.'

At the same time, the German Navy began to consider

seriously the use of similar tactics themselves, especially to protect their U-boats. On 14th May, during a conference with the Führer, Admiral Dönitz reported:

> We are at present facing the greatest crisis in submarine warfare, since the enemy is for the first time making fighting impossible and causing us heavy losses, by means of new location devices.

The U-boats had to be equipped with a number of devices to reduce their vulnerability. A *Naxos-Z* radar detector was already being developed to enable German fighter aircraft to home on the *H2S* transmissions; now a *Naxos-U* detector was commissioned for the submarines, to give warning of the approach of an Allied aircraft fitted with centimetric radar. Staatsrat Dr Hans Plendl devoted considerable theoretical research to the possibility of making U-boats 'invisible' to centimetric radar by coating them with a special material.[7] And, emulating the 'Window' principle, the U-boats were provided with strings of radar decoy buoys to release in an effort to gain relief from their tormentors. The decoy buoys – code-named *Aphrodite* – consisted of a small hydrogen-filled balloon connected to a sea-anchor by about sixty feet of thin wire. Three strips of aluminium foil about thirteen feet long were attached to the wire at intervals of about twenty-seven feet. In theory, the decoys should have produced submarine-like echoes on the patrol aircraft's radar sets, but in practice they were only ever seen on the radar sets of surface vessels. The tactic achieved little success in consequence, but, for the record, yet another service had used 'Window' before the RAF. The idea was not entirely dead in the German air force either. On 15th June, Colonel von Lossberg raised the matter towards the end of a Berlin Air Ministry conference: 'I wanted to ask Herr Feld-

7. In December 1943, Plendl reported to Himmler that he had devised a process promising a 50 per cent reduction in a U-boat's radar reflection, called '*Netzhemd*' – 'Aertex shirt'. The research was generally referred to as 'black U-boat' work. In spite of Plendl's optimistic report, and the capture of full information on his research after the war, there is still no such thing as an 'invisible' submarine.

marschall for an interview to discuss the report on *Düppel* . . .' Colonel Pasewaldt also asked to see him: 'There is a very grave situation arising now, about which I must inform Herr Feldmarschall without delay. I should like to ask for time as well.' Field-Marshal Milch said: 'Let's say tomorrow afternoon at 3 p.m.' Unfortunately no record has survived of what was then discussed.

By this time, the orthodox German system of nightfighter defence was at its peak efficiency: the line of nightfighter 'boxes' running from Denmark down to the Swiss frontier was claiming an increasing number of Bomber Command's aircraft, and many more aircraft were falling victim to the radar-directed flak batteries ringing the main cities of the Ruhr. As the Battle of the Ruhr passed its climax, the loss rate in the bombers continued to mount each night, and morale in the Command began to sag. It was clearly time for the final offensive on the 'Kammhuber line's' *Würzburg* system to begin.

For one more month the superb ground-controlled interception system remained unimpeded in its operation. The operations on the night of 22nd June afford a good example: some minutes before 1 a.m., the young German Second-Lieutenant Heinz-Wolfgang Schnaufer was scrambled from the Saint-Trond airfield in Belgium, in his Messerschmitt 110. The *Freya* early-warning radar had detected a large force of RAF bombers crossing the North Sea. Schnaufer was ordered to the box code-named '*Meise*', fifteen miles north-east of Brussels. The RAF target that night was Krefeld, in the Ruhr, and the route should have passed well to the east of '*Meise*', but at 1.20 a.m. Schnaufer was informed by radio of a lone bomber – far off course – approaching the box from the west. On the ground, the men of No. 13/211 signals company manning '*Meise*' were already tracking Schnaufer's Messerschmitt with one Giant *Würzburg* radar; now the other swung round and began sweeping the night sky, looking for the raider. The hand-over from the *Freya* early-warning system went without a hitch, and by 1.26 a.m. the flight path followed by the unsuspecting British crew was already appearing as a series of co-ordinates on the fighter-control's grid, and as a red spot of light went across the frosted screen of '*Meise*'s' Seeburg table, Second-Lieutenant Kühnel, the fighter-control officer of the '*Meise*' signals company, guided Schnaufer over the radio-telephone into position for a 'parallel head-on' interception. This form of attack, designed to bring the fighter into contact with its quarry at the greatest possible range from the ground radar, allowed the maximum room for error with the limited cover of the precision radars. Schnaufer's orders were to fly straight towards the bomber then, just before the two air-craft crossed, turn through a half circle; the night-fighter slid round neatly on to the tail of the bomber – a perfect interception. In the rear of the Messerschmitt Second Lieu-tenant Baro, the radar operator, observed a small hump of light rise up from the flickering base line of his screen: an enemy aircraft, range 2,700 yards. No need for further instructions from the ground, unless things went wrong. Baro passed Schnaufer a running commentary on the bomber's position until 1.30 a.m. when, in Schnaufer's words:

> I recognized at 500 yards above and to the right a Short Stirling and succeeded in getting in an attack on the violently weaving enemy aircraft. It caught fire in the fuse-lage and wings, and carried on still blazing. Then it went into a dive, and crashed two miles north-east of Aerschot.

At first light, the fighter-controller drove out to inspect the bomber's wreckage to verify Schnaufer's claim: 'There was a crew of seven,' he reported, 'all of whom were lying dead in the wreckage.'

This was the defensive weapon which General Josef Kammhuber had forged and tempered. But it was entirely dependent for its success on the Giant *Würzburg* and *Lich-tenstein* radar systems, and the Germans were soon to realize that these were all too vulnerable to interference by the enemy.

*

For many German air force officers, there was something *too* disciplined and systematic about Kammhuber's system, and early in July the first cracks began to appear in the rigidly controlled night-fighting organization. As the tactic which subsequently emerged was to play an important part in the immediate German countermeasures to the RAF's first use of 'Window', it must be looked at in some detail.

Six months before, a former bomber pilot, Major Hajo Herrmann, had suggested to Kammhuber that in view of the slow output of two-seater night-fighters, the force should employ the single-seater day-fighters in the night defence of the Reich as well. They would combat the enemy bombers over the actual target areas, where they were illuminated by the searchlights, the vast conflagrations and the Pathfinder flares. The single-seater aircraft would fight over the Bomber Command target for as long as they could, and then land at the nearest airfield to refuel. Kammhuber had always opposed the operation of fighter aircraft over flak areas, however, and he would not accept the idea. During the RAF attack on Berlin on 1st March – which has already played an important part in this story – Herrmann watched the air battle from the anti-aircraft guns' divisional headquarters. The flak commander agreed that he could restrict the ceiling of fire to a certain altitude over Berlin – above which ceiling Herrmann would operate with a picked bunch of ex-bomber pilots, flying single-seater aircraft. From 8th April onwards, Herrmann pressed Colonel-General Weise to start experimental operations on these lines, and Weise, who was Kammhuber's superior, allowed him to begin. The first such operation took place on the 20th during an attack on Stettin, which was coupled with a Mosquito raid on Berlin; but this time the *Freya* warning came too late and Herrmann's intrepid band could not reach the Mosquitoes' altitude in time.

Realizing that the weight of the attack was now on the Ruhr, he moved to the west and took up station near the Ruhr. He secured the co-operation of the local AA commander, General Hintze, to try out his theories over Essen and Duisburg. But the next big attack by the RAF was on

Cologne, outside Hintze's 4th Flak Division area, on 3rd
July. Unaware of this, Herrmann's twelve men took off
from München-Gladbach in five Focke-Wulf 190's and seven
Messerschmitt 109s's, and soon they were in a running fight
all the way to the target. Herrmann had not reached any
agreement about a ceiling of fire with the Cologne flak commander, General Burckhardt, and soon the weight of the AA
barrage forced them to break off their engagement – but
not before the twelve aircraft had shot down twelve of the
thirty-two aircraft the RAF lost that night.

On the following morning there was a telegram of con-
gratulations from Weise, and before the day was over he
was telephoned by Göring himself: the major was ordered
to report to the Commander-in-Chief at once at Karinhall
– Göring's country retreat near Berlin. That evening he ex-
pounded his full theory to Göring: all available fighters
allocated to the system would assemble over radio beacons
when the approach of bomber formations was signalled.
From there they would be hurled *en masse* into the bomber
stream right over the target; they would pursue the bombers
until the limit of their endurance, and then refuel at the
nearest airstrip. The new tactic was given the apt code-name
'Wild Boar'.[8] Herrmann was given permission to form a
fighter wing, comprising three squadrons of Messerschmitt
and Focke-Wulf single-seat fighters.

Herrmann personally addressed a top-level conference
in Berlin on 6th July:

I would like to stress the following. In the area of one
flak division in the Ruhr, where the weather is moderately
clear, one can reckon that an average of 80 to 140 enemy
bombers are picked up and held by searchlights for longer
than two minutes, during an air raid. The demand I am
making of my crews is that every aircraft held for more
than two minutes is shot down. I would go so far as to
say that if the British continue their attacks in weather
conditions like these, they can lose an extra eighty air-

8. *Wilde Sau.*

craft every night, if I am provided with the necessary personnel.

Field-Marshal Milch agreed vehemently: how often they had gazed at some British bomber trapped in a searchlight cone, and longed for a German single-seater to be up there closing on its tail. Now it was beginning to come true. He asked Herrmann: 'Have you enough crews?' Herrmann replied: '*Jawohl!* There are 120 crews there!' 'And how many planes do you have?' 'In my experimental unit I now have fifteen aircraft, and fifteen more have just been allocated to me.' Milch ordered that all the aircraft needed by Herrmann were to be made immediately available. Somewhat emotionally, General Vorwald exclaimed: 'At last the spell is broken!'

At no point in the new tactics would Herrmann's force have to rely on electronic gear other than the old-fashioned early-warning radar and the radio beacons they would need to find their way around. In consequence the new 'Wild Boar' tactic would effectively short-circuit the RAF's impending introduction of 'Window' jamming of *Würzburg* and *Lichtenstein*. Herrmann stressed that he was not advocating solely one- or two-seater night-fighting, so much as a new concept: the massive and unified operation of all available forces, hurled *en masse* against the Bomber Command formation once its real target had become known, 'whether it be Cologne, Hamburg or Berlin.' Major Herrmann had promised Göring that his force could be ready in three months – the beginning of October. In fact it was to be thrown into the fray very much earlier than that.

All this was anathema to General Kammhuber.

A few days later, at a further Berlin air conference on 16th July, he drew particular attention to the importance of the Giant *Würzburg* chain:

Where the bomber stream enters on a narrow front, for example in the Ruhr and over the Heligoland Bight where there is very little early warning, it is vital that the few 'boxes' that are affected are 100 per cent operational. Most unfortunately, it keeps happening that the most important 'boxes' break down, like No. 5 'box' for instance. The loss of one 'box' during one night brings the number of kills right down.

Together with Martini, he urgently asked for *Freya* radar production to be doubled or even trebled. One of the night-fighter control experts, Günthner, described in detail a new system being developed for funnelling night-fighters right into the bomber stream, using radio beams similar to those used over Britain in 1941. He requested the provision within the next two months of thirty night-fighters to exploit the new system – code-named 'Y-control' – to the full: 'Every week a city is being smashed, so we must act fast', said Günthner. And later he added: 'If I think of how we can add eighty kills to the twenty we are scoring now, that will mean they drop their attacks altogether.' Milch mockingly answered: 'Very well, Günthner, we'll do that. Let's appoint Günthner the Third General of Night-Fighters – he's got his head screwed on.' The idea put forward by Major Streib, a leading night-fighter ace, was that the night-fighters should be fed into the RAF bomber stream somewhere near the Scheldt estuary, and pick the bombers off one by one. Field-Marshal Milch said: 'So they keep up with them, shoot them down one after the other and keep asking: "And the next one please!"' To overcome the problem of identity, the fighter pilots were to be instructed not to attack anything without four-engines. 'Those are the worst ones anyway,' said Günthner.

In general the conference had a light-hearted tone, for with Herrmann's new system they felt sure they could bring their nightly scores even higher. Colonel von Lossberg was talking about a new night-fighter device under development which would enable them to home on to the fighter-attack warning device fitted to the RAF bombers, code-named 'Monica'. But Milch ended the discussion on a serious note: 'The best that can be achieved by us is this,' he said. 'The enemy cease to operate against us altogether. Until we can get things in order, however, the best we can hope is to make his business a messy one. In that case, there is only one worry for us, and that is that in some way he again catches us on the hop with some radar-trickery, and we have to start trotting after him again.'

That day, 16th July, a Reich Radar Research Authority[9] was founded to supervise an expansion of the German radar and high-frequency electronics industries, and organize all fundamental research. Kammhuber, Milch and Plendl agreed to hold regular meetings to sort the radar problems out. Soon, Plendl had three thousand scientists working under him.

Three weeks before, Sir Charles Portal had brought up the question of the use of 'Window' to jam the German radar system, at a Staff Conference on 23rd June. By now the opposition to the employment of this tactic was crumbling, and even Lord Cherwell, who had been the most formidable member of the opposing school, felt that 'on the whole the time is rapidly approaching when we should allow it to be used.' Mr Churchill gave his final permission to 'open the Window' as soon as possible. The invasion of Sicily was planned to open on 10th July: once this obstacle had been removed Bomber Command could use 'Window'. Sir Arthur Harris was directed to prepare his force to employ the aluminium foil strips as from the first day of July, depending on the progress of the ground forces' operations in Sicily.

In the meantime, Wing-Commander Jackson's 'Window' Panel had set up the complex machinery necessary for the production of vast quantities of aluminium foil: at least four hundred tons of 'Window' – about 1,000 million strips – would be required each month. Since late 1942, certain selected companies had been working at high pressure turning out the 'Window' foil: stiffened by black paper backing, it was guillotined into strips 30 centimetres long by 1·5

9. *Reichsstelle für Hochfrequenz-Forschung.*

centimetres wide; the shiny front was layered with lamp-black so that the clouds of 'Window' should not show up in the searchlight beams. Two thousand strips held together by an elastic band were enough to give an echo similar to a heavy bomber: each unit cost about fourpence.

For maximum effect, Bomber Command would have to launch a whole series of very heavy attacks supported by 'Window' before the German defences could recover their balance. Both Harris and Saundby have referred in this context to the example of the introduction of the tank in the First World War: the tank had been introduced prematurely, and the Bomber Command leaders were anxious that there should be no parallel with 'Window'. It had only been in May 1943 that the firms concerned had been able to meet Bomber Command's needs – a factor which has often been ignored when the 'Window' controversy has been examined in the past.

The final conference to discuss the use of the new tactic was held on 15th July – the day before Field-Marshal Milch had expressed his fears about possible new 'radar-trickery' by the British. By this time, the Allied troops had established a firm foothold in Sicily, but Mr Herbert Morrison, Minister for Home Security, was still worried about the German power to retaliate using the same tactic. He asked whether this power could not be reduced still further by bombing the enemy's airfields. Sir Charles Portal said that this would be a waste of effort, since the German bomber force was 'weak, badly trained and over-extended'. Mr Churchill affirmed that he was personally prepared to take the responsibility for the introduction of 'Window'. Morrison withdrew his objection and the last barrier was removed. The conference decided that Bomber Command should use 'Window' to cover their attacks from 23rd July.

7

Gomorrah and After

The attacks on Hamburg have affected the morale of the people. Unless we evolve a means of defeating these terror-raids soon, an extremely difficult situation will arise.

– FIELD-MARSHAL MILCH, 30th July 1943

Thus fate was hastening the controversy to its inevitable conclusion. For sound tactical reasons, Sir Arthur Harris decided that the city which should be the target for the forthcoming operation – operation GOMORRAH – would be Hamburg. The summer nights were short and long overland flights to the same target night after night would be impossible; but Hamburg was on the very seaboard, and its distinctive waterfront and dockyard areas showed up unmistakably on the Pathfinders' *H2S* radar equipment. Harris warned that the 'Battle of Hamburg' could not be won in a single night: 'It is estimated that at least 10,000 tons of bombs will have to be dropped to complete the process of elimination. To achieve the maximum effect of air bombardment this city should be subjected to sustained attack.'

Hamburg was the second largest city in Germany and the largest port in Europe. In 1940 the battleship *Bismarck* had taken shape there in the sprawling yards of Blohm & Voss; but now the three main shipyards were turning out U-boats as quickly as they could. Clearly Hamburg was a military target of the highest importance. By the third week in July 1943, the city had been raided on ninety-eight occasions. Her defences were strong: Hamburg was ringed by no less than 54 heavy flak batteries, 22 searchlight batteries and three

smoke-generating units. Twenty radar ground-control
'boxes' served by six major night-fighter airfields covered
the approaches to the port, and previous operations had in
consequence cost the RAF many casualties.

On the afternoon of 24th July, almost every operational
crew in Bomber Command crowded the bomber stations'
operations blocks for the briefing on the night's raid. The
target announcement raised a disgruntled murmur, as it
had a formidable reputation. The men could see their route
before them, lined in ribbons pinned across the wall maps:
assemble over the North Sea, fly on an easterly course
clear of the German defences, then turn south-east and
attack; then they were to clear the target area on a reciprocal heading.

Next came the plan of attack: zero-hour was to be 1 a.m.,
three minutes before which twenty Pathfinder aircraft would
release yellow target indicators and illuminator-flares blindly
on the indications afforded by *H2S* alone. One minute later,
eight crews were to mark the target visually using red target
indicators, having been guided in by the flares already dropped. The main force would start bombing at two minutes
past 1 a.m. and the last aircraft was to be clear of the target
area forty-eight minutes later.

The crews had been briefed earlier on how and where
'Window' tactics were to be used: from the meridian of
$8\frac{1}{2}$ degrees east on the target-bound flight until 8 degrees
east on the homeward flight the crews were to drop one
'heavy-bomber' bundle every minute. At the end of the
briefing, a carefully scripted announcement from Bomber
Command headquarters was read to the crews:

> Tonight you are going to use a new and simple counter-
> measure, 'Window', to protect yourselves against the
> German defence system. 'Window' consists of dropping
> packets of metal strips which produce almost the same
> reactions on RDF as do your aircraft. The German de-
> fences will, therefore, become confused and you should
> stand a good chance of getting through unscathed while
> their attention is wasted on the packets of 'Window'.

The badge of No. 360 Squadron, the Royal Air Force's specialist electronic countermeasures unit, on which the author served for some years. The centrepiece, resting on a trident, is the Melese Laodamia Druce moth which lives in Central America. When it hears acoustic radar signals from a bat hunting for its prey, this creature replies with signals of its own which effectively jam the radar. These moths have been in the radar jamming game for some millions of years, and probably started it all.

The first photographic evidence to reach Britain showing that the Germans, too, had radar equipment in service: the *Graf Spee* pictured on 19th December 1939 after she had been scuttled off Montevideo; the radar aerials can be seen on the front of the main director tower.

Heinkel III's of the pathfinder unit *Kampfgruppe 100*. On the nose of the aircraft nearest the camera can be seen the 'Viking Ship' emblem. Each of the unit's aircraft carried three aerials on the rear fuselage: one in the front for normal radio communications and two in the rear for the *X-Gerät* beam flying equipment.

The *Würzburg* precision radar was the object of considerable interest from RAF Intelligence; the examples depicted served flak batteries.

Early on the morning of 28th February 1942, British paratroopers seized and dismantled the *Würzburg* radar at St Bruneval, near Le Havre. The raid resulted from a reconnaissance photograph taken by a Spitfire pilot nearly three months earlier.

One of those who took part in the work of dismantling the German set was Flight Sergeant Cox (below, second from the left), an RAF radar technician.

The dramatic low altitude reconnaissance photograph of the Giant *Würzburg* radar at Domburg on the Dutch island of Walcheren taken on 2nd May 1942. One of the operators stands helplessly beside his set to become a human yardstick when British Intelligence officers analysed the picture.

A typical *Himmelbett* station with a *Freya* radar (centre) for long range search and two Giant *Würzburg* sets to track the movements of the enemy bomber and the intercepting night fighter.

Dr R V Jones headed the
Scientific Intelligence
section at the Air
Ministry in London
which pieced together
details of the German
radar system.

Engineer Colonel Dietrich
Schwenke headed the
German Intelligence
efforts to probe the
secrets of captured Allied
equipment, including the
revolutionary *H2S* radar.

General Josef
Kammhuber the architect
of the *Himmelbett*
system of controlled
night-flying.

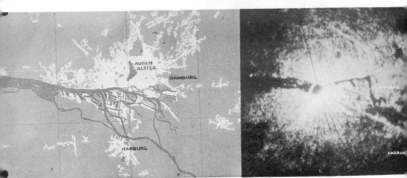

H2S picture of Hamburg (right) compared with a map of the same area. The wide estuary of the River Elbe, pointing at the city from the west, served as a prominent navigational feature on radar.

'Window' was a secret that both sides took great pains to keep to themselves. Dr D A Jackson (right), played a major part in the work to bring 'Window' to the point where it could be used to reduce Bomber Command's losses. Lord Cherwell (left), Mr Churchill's scientific adviser, opposed the use of 'Window' until the summer of 1943. Then the clouds of metallic foil strips were used with great effect to wreck General Kammhuber's elaborate defensive system (below).

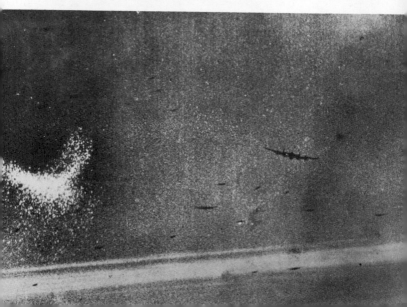

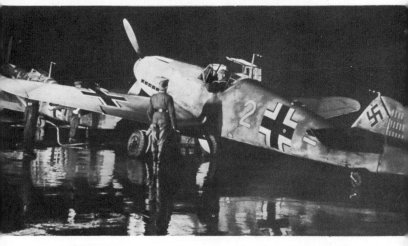

As an initial counter to 'Window' the *Luftwaffe* sent single-engined day-fighters to engage the night bombers: 'Wild Boar' Messerschmitt 109's, their pilots at cockpit readiness, seen awaiting the order to take off.

Major Hajo Herrmann, centre, seen with Goering during an inspection of 'Wild Boar' crews. Herrmann was the instigator of the 'Wild Boar' tactics.

Following the collapse of General Kammhuber's defensive system as a result of 'Window' jamming, Kammhuber was removed from the command of the night-fighter force. His place was taken by General Josef Schmid (centre).

Two German pilots who amassed large scores using 'Tame Boar' tactics: Prince Heinrich zu Sayn-Wittgenstein (left) and Heinz-Wolfgang Schnaufer.

In the autumn and winter of 1943 the Royal Air Force stepped up its support for the night raiders. No. 101 Squadron's Lancasters carried high powered ABC radio jammers; two of the aerials used by this device can be seen on the upper fuselage. The objects falling away from the bomber were incendiary bombs.

From the radio station at Kingsdown in Kent, RAF operators broadcast false orders to German night-fighter crews.

The British Intelligence service's attempts to discover the details of the radar sets fitted to the German night-fighters was an exciting and sometimes dangerous operation. A Junkers 88 with the clumsy Lichtenstein aerials.

A remarkably clear camera-gun picture taken early in 1944 - a Junkers 88 weaving in a vain attempt to shake off an American long-range fighter. This photograph, revealing an unusual structure on the nose, provided the first clue as to the nature of the latest German radar, *SN-2*.

Dr Robert Cockburn and his jamming section at Malvern devised the 'ghost fleet' feint to support the Normandy landings.

The feint was rehearsed against captured radar sets at Tantallon Castle on the Firth of Forth; the *Würzburg* can be seen in the foreground.

From the autumn of 1944 German night-fighters had to be equipped to avoid British intruders, as well as to shoot down bombers. The Junkers 88G carried *SN-2* radar, with nose aerials to cover the forward hemisphere. The blister on top of the cockpit housed the aerial for *Naxos,* which picked up signals from both bombers' *H2S* equipment and the British night-fighters' radars. The main armament was housed in a bulge under the fuselage.

A backwards pointing machine-gun was fitted to the rear of the cockpit for self-defence. Note the pair of upwards-firing cannon.

Air Vice Marshal E B Addison, the commander of No. 100 Group.

A specialist jamming Liberator of No. 233 Squadron, one of the units belonging to No. 100 (Bomber Support) Group.

B-17 Flying Fortresses under fire from accurate, radar-laid, flak.

The 'Carpet' jamming transmitter, which provided USAAF heavy bombers with a useful degree of protection from radar-laid flak.

Since the end of the Second World War both power blocks have regularly sent aircraft to probe the defences of potential enemies and record radar signals. The Lockheed U-2 reconnaissance aircraft, whose overflights deep into Russian territory sparked off a major international incident.

A Royal Air Force Phantom fighter maintains station on a Russian Tupolev 20 reconnaissance aircraft (NATO code-name 'Bear') off the coast of Scotland.

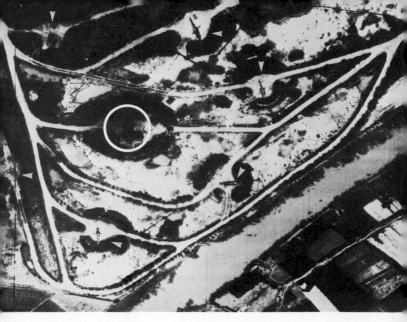

Overall view of an *SA-2* surface-to-air missile battery, photographed in North Vietnam. The control radar vehicle (NATO code-name *Fansong*, circled) is in the centre of the firing complex and the six launchers for the missiles (NATO code-named 'Guideline', arrowed) are spaced evenly round it at a distance of about 80 yards.

Model of a *Fansong* radar truck (top right). The vertical trough aerial on the right radiates a fan-shaped beam which is swept up and down to measure the elevation of the target; the horizontal trough aerial in the centre radiates a similar beam which is swept from side to side to measure the azimuth of the target. Scanning is electronic, that is to say that the beams sweep without the scanners moving, once the target is near the centre of their field of view. The dish-shaped aerial on the left radiates command guidance signals to the missiles after launch.

Close-up of a 'Guideline' missile on its launcher (bottom right). This weapon has a maximum speed of about Mach 3.5 and a maximum effective slant range of about 25 miles.

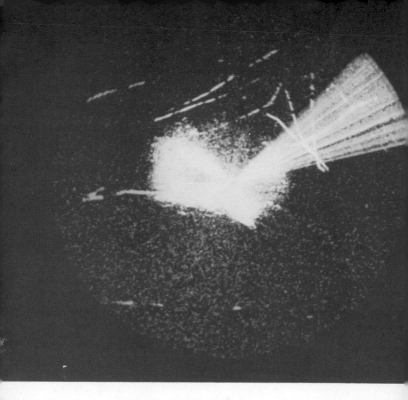

A radar screen showing a typical 'spoke' caused by transmitted noise jamming from a single aircraft. The jamming 'spoke' originates at the centre of the screen and extends to maximum range, concealing aircraft on the part of the screen it covers; the width of the 'spoke' is proportional to the amount of jamming power radiated on the radar's frequency. The main effect of this type of jamming is to deny the radar operator range information on the aircraft radiating it, though he can still get a rough idea of the bearing of a single jamming aircraft; to provide maximum confusion, several widely-spaced aircraft will jam simultaneously. The broken lines on the screen are 'Chaff' trails laid by other aircraft.

Typical 'Chaff' for use against modern 3cm wave-length radars (top right).

'Rope' used to jam radars working on long wavelengths. As it falls clear of the aircraft the package releases the reel of foil. Suspended from the small cardboard 'parachute' the latter unwinds to leave a slowly falling strip hanging vertically in the sky (bottom right).

A 'Quail' radar decoy missile falling clear of a B-52D long-range bomber. The decoy carries special reflectors to give it the same appearance on enemy radar as the B-52 itself.

Electronic Warfare Officer in his position in a B-52G.

Eight nuclear-headed *SRAM* air-to-surface missiles, mounted on a rotary launcher in the bomb bay of a B-52. This multi-attack capability enables the bomber to clear its own path through the defences, if necessary.

The B-52H, final production version of the aircraft which has formed the backbone of the USAF long-range bomber force for nearly two decades. The bulges under each side of the nose house the viewing systems for the forward-looking infra-red and low-light-level television devices; these enable the crew to fly close to the ground with the cabin blacked-out, to prevent their being blinded by the flash from exploding nuclear weapons. To the rear of the viewing blisters are aerials for some of the jamming transmitters.

A Royal Air Force Vulcan bomber, showing the bulged tail cone carrying the jamming equipment.

A jamming transmitter, fitted in a pressurised canister, being readied for installation into a Vulcan.

A Canberra Mark 17 electronic countermeasures aircraft of the Royal Air Force.

Ground crewmen loading a 'Chaff' dispenser into the wing-pod of a Canberra Mark 17.

A fighter-bomber goes to war. A USAF F-4 Phantom carrying an up-to-date war load: on the outer wing stations, two fuel tanks; on the inner wing stations, two 2,000 pound laser guided bombs; under the rear fuselage, two 'Sparrow' semi-active radar homing air-to-air missiles for self-protection; and under the centre fuselage two ALQ-71 or ALQ-87 electronic jamming pods.

Close-up of an ALQ-87 jamming pod, fitted to the inner wing station of a Phantom; four radiating aerials, two large and two small, extend from the underside of the pod.

SA-2 surface-to-air missiles fired at US aircraft over North Vietnam. The evasive manoeuvres and jamming by the target aircraft have made them miss by a wide margin.

The US Navy's Grumman EA-6B Prowler carries its ten jammers in five streamlined pods fitted under the wings and the fuselage, with the associated receiver equipment in the fairing on the top of the fin.

A USAF EB-66E, its fuselage bristling with jamming aerials. Laying out 'Chaff' trails or providing jamming support within range of the ground defences can be a hazardous business; so remotely controlled electronic warfare drones have been developed for this purpose, like this Ryan AQM-34 (below) seen under the wing of its DC-130 launching aircraft.

One interesting innovation during the Vietnam war was the formation of special 'Wild Weasel' defence-suppression units, to engage enemy anti-aircraft gun and missile units by launching radiation-homing missiles at their control radars. This 'Wild Weasel' F-105G Thunderchief belongs to the 57th Tactical Fighter Wing.

Close-up of the Anglo-French 'Martel' radiation-homing missile.

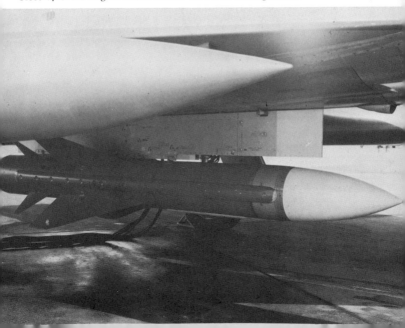

An interesting case of the use of design features to render an aircraft less vulnerable to enemy missiles can be seen on the Fairchild A-10 attack aircraft. The engines have been mounted high on the rear fuselage where, from aspects such as this one, the unusual twin-finned tail screens the hot exhausts from ground launched infra-red homing missiles.

A shape of things to come? A Grumman design for a high-altitude penetrator intended to have the smallest possible radar signature. Note the unusual shape of the wings and fuselage, to reduce to a minimum the strength of the reflected radar signals. The engine intake and exhaust pipe, both of which can be great reflectors of signals, have been positioned above the fuselage out of view of ground radars when the aircraft is at high altitude.

Approximately 80 aircraft at Gustav-Caesar 5, course east, altitude 19,000 feet.

A small spot of light was jerked across the darkened Situation Map at the Third Fighter Division's headquarters at Arnhem-Deelen in Holland; it came to rest at position *GC-5* on the German fighter grid, just to the north of Ipswich.[1] Soon the spot of light was joined by others as more and more bombers appeared over the radar horizon. The attack was obviously going to be a big one.

At a quarter-past midnight, the leading aircraft turned south-east into the Heligoland Bight, track one-one-seven. The aircraft's bomb-aimers crawled back into the rear of the fuselage where the 'Window' bundles were waiting stacked around the flare chute. At twenty-five minutes past midnight, the first bombers were passing the $8\frac{1}{2}$-degree meridian. One bundle per minute, the aluminium foil was dropped into the black air beneath each bomber. At the main force's altitude, the temperature was minus 20 degrees Centigrade, and even holding the torch and stopwatch was an uncomfortable burden for the bomb-aimers. At twenty to one, the leading aircraft were crossing the enemy coast. The flight-engineers crawled back to the flare-chute and relieved the bomb-aimers, who would soon have other tasks on hand. Ten minutes later, the bombers turned on to their attack heading, one-six-six degrees True. In front of the *H2S* radar operators, the large echo that showed Hamburg was gradually moving in towards the centre of the screen. At three minutes to one, right on time, the first yellow target markers were released. The Battle of Hamburg had begun.

In Hamburg the sirens had already sounded twice during the day – once just after midday and again at 9.30 p.m. In both cases it had been a false alarm and the All Clear had sounded soon after. At nineteen minutes past midnight, as the vanguard of the bomber stream came within 130 miles of the city, a restricted warning went out to civil defence

1. The bombers had not reached 19,000 feet at this position; possibly there were freak radar conditions.

headquarters, industrial buildings and hospitals: Air Danger 30 – an attack was possible in the next thirty minutes. Air Danger 30 was the cue for the local air defences to come to life: at flak sites all round Hamburg, men tumbled out of rest rooms and began to ready their weapons. Covers came off, barrels rose skywards and swung round to the north-west, shells were fused and radar sets warmed up. From the fighter-control bunker 'Socrates', built into a cliff-face twelve miles west of Hamburg, the duty officer of the Second Night-Fighter Division ordered crews to cockpit-readiness. At the fighter airfields at Stade, Vechta and Wittmundhaven, Luneburg, Jagel and Kastrup, the aircrews hastened to comply.

At twenty-four minutes past midnight, the restricted alert was amended to Air Danger 15. Seven minutes later, the first public *Fliegeralarm* was sounded – a series of two-second siren tones. For a full minute, the rising and falling siren blast echoed dis-synchronously across the city from a hundred sirens, from Blankenese in the west to Wandsbek in the east, from Langenhorn in the north to Harburg in the heart of the docklands in the south. But the inhabitants of Hamburg were not alone as they ran for their shelters: nine Mosquitoes were dropping target markers and a few small bombs to cause diversions over Lübeck, Kiel and Bremen, and in each city the full *Fliegeralarm* sent people scurrying for cover.

The first report that something out of the ordinary was happening came from the '*Hummer*' radar 'box' sited on the island of Heligoland. Nor was it surprising, for the bomber stream would by now be appearing together with the 'Window' on the *Würzburg* screens: as each 'heavy bomber' cloud remained effective for fifteen minutes before the strips dispersed, the 746 aircraft were returning radar echoes like those from a force of eleven thousand bombers! The '*Hummer*' station's radar operators complained that they were 'disturbed by many apparent point-targets looking like air-craft, either stationary or slow moving. The picking-up of genuine aircraft is made extremely difficult. Once they have

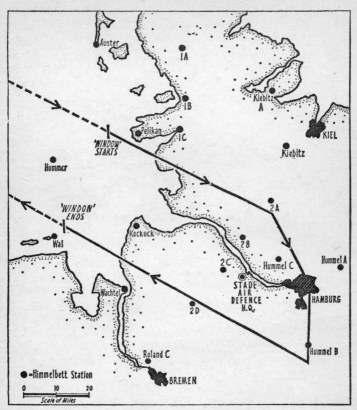

The Battle of Hamburg

been picked up it is possible to follow them, but only with difficulty.' The '*Auster*' station located on the southern tip of Sylt had similar trouble. So, in turn, did the rest of the radar stations sited around Hamburg proper.

Circling over their appointed radio beacons, the German night-fighter crews waited with growing impatience for instructions from ground control. But the men on the ground could not help them. Soon the ether was thick with confused appeals and exclamations: 'The enemy are reproducing themselves.' 'It is impossible – too many hostiles.' 'Wait a

while. There are many more hostiles.' 'I cannot control
you.' 'Try without your ground control . . .'

In the fighter-aircraft themselves, the same scenes of confusion were being reproduced as their *Lichtenstein* air-to-air
radar sets were swamped by the drifting clouds of aluminium foil. What happened is best described by one of the
German pilots who saw action this night:

At 5,000 metres my radio operator announced the first
enemy on his *Lichtenstein*. I was delighted. I swung round
on to the bearing, in the direction of the Ruhr, for in this
way I was bound to approach the stream. Facius proceeded to report three or four targets on his screens. I hoped
that I should have enough ammunition to deal with them!

Then Facius shouted: 'Tommy flying towards us at
a great speed. Distance decreasing . . . 2,000 metres . . .
1,500 . . . 1,000 . . . 500 . . .'

I was speechless. Facius already had a new target. 'Perhaps it was a German night-fighter on a westerly course,'
I said to myself, and made for the next bomber. It was
not long before Facius shouted again: 'Bomber coming
for us at a Hell of a speed. 2,000 . . . 1,000 . . . 500 . . .
he's gone!'

'You're crackers, Facius,' I said jokingly. But soon I
lost my sense of humour, for this crazy performance was
repeated a score of times.[2]

Some of the fighter-pilots did get 'kills', but they were few
and far between.

One of the successful ones was Sergeant Hans Meissner,
an Me 110 pilot with *NJG 3*'s second squadron. He too had
made several attacks on 'Window' clouds before his radar
operator, Corporal Josef Krinner, noticed that one 'blip'
on his screen seemed to be almost stationary while all the
others moved.[3] Krinner directed his pilot towards it, and the

2. Wilhelm Johnen: *Duel Under the Stars* (William Kimber, 1957).

3. Moved, that is, relative to the Messerschmitt; the aircraft producing the 'blip' was moving on a similar course at roughly the
same speed, so – on the radar screen – it appeared to be stationary.

latter caught sight of a four-engined aircraft, a Stirling. Meissner made one devastating firing pass and the raider burst into flames and dived into the ground.

Another Me 110 of *NJG 3*'s second squadron caught a Halifax over southern Denmark, well off course and outside the protective umbrella of 'Window'; it too received no quarter.

The Pathfinders had done their work well, and at two minutes past one the first wave of bombers arrived over the city – 110 Lancasters of Nos. 1 and 5 Groups. The first thing that struck them was the air of unreality: the 'blue' beams of the dreaded Master searchlights – usually bolt upright before tilting swiftly over to trap some unfortunate bomber – were wandering all over the sky. Indeed, all the searchlights seemed to be groping blindly for the enemy. Where beams did cross, others would quickly join them, and as many as thirty or forty beams would build up to form a cone – on nothing.

'We could hear excited voices coming from the radar cabin,' recalled one of the crews of the 88-millimetre flak battery *1/607* on the Harburg hills. 'There was a wild display of flickers on their cathode-ray tubes, clouding the whole of the screen. The battery-commander, Lieutenant Eckhoff, telephoned at once to the nearest battery to us. The *Würzburg* there had been put out of action too.' On telephoning the divisional operations room, the battery commander learned that every flak radar set in Hamburg was out of action in the same way.[4] The gunners abandoned predicted fire and loosed off round after round into a box in the sky centred on the bombers' predicted release point. But there was a lot of sky around each aircraft, and the chances of a shell bursting at just the right point in time and space were remote indeed.

In the past, height had always meant safety for the bombers. This was no longer the case, for it was the bombers in the van and at the highest altitude that received least

4. David Irving: *Und Deutschlands Städte Starben Nicht* (Zurich, 1963).

protection from the 'Window' tactic. Some of the Lancasters attacked from as high as 23,000 feet, and these showed up on the *Würzburg* screens as clearly as a tree stands out of a low mist. No. 103 Squadron, whose Lancasters were the highest fliers of all, lost three aircraft this night; of the Stirlings, which had the worst height performance, only three were lost. Altogether Bomber Command lost only twelve aircraft (1·5 per cent) of the large attacking force – four Lancasters, four Halifaxes, three Stirlings and one Wellington; probably nine fell to night-fighters and three to the guns. Normally the Stirlings suffered far more heavily than the other types.

The new tactic had clearly been a great success. Had the raid cost the six per cent losses normal for a Hamburg raid, Bomber Command would have lost about fifty aircraft from its force this night. So about thirty-five or more had been saved, by the dropping of forty tons of 'Window' – 92 million strips of aluminium foil.

On 25th July the Führer was roused early and told about the damage wrought in Hamburg. He was also told about the new British tactics, during the midday war conference. Of the damage, he said 'You don't have to tell me that. I have seen the photographs already. But listen : this dropping of tinfoil – does that protect these airmen . . .?'

Lieutenant-Colonel Eckhard Christian, Jodl's General Staff officer, told him: 'The *Würzburg* sets are jammed by it. The night-fighters are still partially controlled by the *Würzburg* sets; that is controlled night-fighting, which guides the fighter to the enemy aircraft. The *Würzburg* sets are now just showing targets everywhere, so they don't know any more where to guide the fighters to. The *Freya* system is not affected – they don't pick up this stuff.'

Hitler asked: 'What kind of sets has the British air force got? Do they pick this up or not?' Colonel Christian unhappily replied: 'I cannot say, my Führer. But we assume so.' 'Do they pick it up, or don't they?' insisted Hitler. 'Find out!'

'*Jarwohl*,' said Christian. 'And the Y-equipment which is

being introduced more and more into our night-fighter system is also unaffected by this tinfoil. That might well be the solution for us to get round this thing.' The trouble was, as he told Hitler, that this depended on the equipment situation, and that was slow in coming. The answer was retaliation against England, but how? At present the German bombers were finding difficulty in navigating to London and back: 'And then I have to listen to some simpleton tell me, "Ja, my Führer, if they come from England over to Dortmund they can drop their bombs precisely on factory-buildings 500 metres long and 250 metres wide using their present radio beams." The blockhead! – And we can't find London when it's thirty miles across and only a hundred miles from our coast.'

Hitler's answer was to sign that day a decree ordering the mass production of the rocket later known as the V-2, but that kind of measure would take time.

Reichsmarschall Göring put more urgent measures into effect during the same morning. He telephoned Major Hajo Herrmann, and asked him whether he could not begin operations at once with at least a part of his special wing of 'freelance' single-seater night-fighters: 'It has been very bad in Hamburg,' said Göring. Herrmann, who had no idea of the scale of the disaster, told the Reichsmarschall that he still needed time. Göring, very agitated, said that Herrmann must do something as soon as possible, and Herrmann promised to have at least twelve aircraft operational that night. Some hours later, he received a teleprinter signal from Göring setting out the details of the previous night's blow.

Working feverishly, Herrmann began to set up a special system to enable his single-seater pilots to navigate by night: the anti-aircraft units of each big city were to fire so-called 'light-houses' into the night sky – distinctive combinations of flares hanging on parachutes. He arranged for the searchlight units to co-operate fully with the night-fighters, and in particular to mark the path of a bomber force overhead by laying a horizontal searchlight beam along the ground in the same direction as the bombers.

The system was ready and waiting when RAF Bomber

Command launched its second attack on Hamburg at three minutes to 1 a.m. on 28th July. The 722 bombers involved flew across the city from north-east to south-east this time, and by twelve minutes past one crews running in to bomb saw beneath them a vast carpet of fire, covering almost the whole of the north-eastern quarter of the city. Into this inferno, succeeding aircraft were dropping thousands of incendiary and high-explosive bombs. During the whole of July, less than 1·7 inches of rain had fallen on the city and the previous day had been very hot: the kindling was everywhere. Under the torrent of well-placed incendiary bombs, the fires soon took hold; with the city's water supplies disrupted and the civil defence headquarters already blitzed in the previous raid, the blaze was able to rage unchecked. The conflagration grew into the 'fire storm' for which the Battle of Hamburg was to become famous in later years.

While the bomber crews gazed awe-struck on the holocaust below, the RAF's Wireless Intelligence service in England was picking up details from the German radio traffic which strongly suggested that it had undergone a marked reorganization – as indeed it had. 'Instead of the usual brief instructions as to course and height, ground stations were heard to give something of a running commentary regarding the course and height of the British aircraft and information about their being held in searchlights. The conclusion to be drawn is that the enemy decided to use a system of much looser control of his fighters when interference from "Window" made it necessary. In the traffic there were several direct references to fighters flying without ground control.' The operations by Herrmann and his recently formed fighter wing had not passed unnoticed. While the overall losses from this raid – 17 aircraft – were still remarkably low, the rate was rising again.

Even so, the scenes of terror in the centre of the Hamburg fire-storm area were almost indescribable. Major-General Kehrl, the local civil defence chief, reported:

Children were torn away from their parents' hands by the force of the hurricane and whirled into the fire. People who thought they had escaped fell down, overcome by the devouring force of the heat, and died in an instant. Refugees had to make their way over the dead and the dying. The sick and the infirm had to be left behind by the rescuers as they themselves were in danger of burning.

Further attacks were expected and that afternoon the city's defence commissioner, Gauleiter Kaufmann, appealed to all non-essential civilian personnel to leave Hamburg. They needed no second bidding. Between dawn and dusk, nearly a million civilians, many of them swathed in bandages, streamed out of the city limits; as night fell, virtually the only people left were those employed in the fire-fighting and defence services.

But still Hamburg's ordeal was not over. Yet a third attack was mounted on Hamburg by Bomber Command: as the Pathfinders ran in to begin their marking at twenty-three minutes to 1 a.m. on 30th July, they could see the fires still burning from two days before. Now Hamburg was wide open – her water mains were smashed, her roads were blocked. But if her guns were ineffective, Herrmann's men were having some success. Bomber Command's operational research section reported: 'The number of searchlights had been greatly increased both *en route* and in the target area. An outer belt stretched in a semicircle round the town from the north-east to the south-west, and inside this other searchlights apparently acted as fighter guides, sometimes exposing horizontally for track indicating and to silhouette attacking bombers.' The bombers reported that the German fighters intercepted them mostly in the height band from 17,000 to 20,000 feet, and this in itself was an unusual feature; again the same general 'running commentary' on the movements of the bomber stream, rather than of individual aircraft, had been broadcast to the German night-fighters: 'Some seem to have landed, refuelled and taken off again – an unusual procedure, but one which would be encouraged in a freelance system.' Not long after, the wireless monitors overheard an even odder feature: some of the call-signs used indicated that the aircraft were day-fighters, not night-fighters at all. During this third attack on Hamburg, only twenty-seven of the 777 bombers dispatched had been lost: but the loss rate – 3·5 per cent – was clearly an increase on what had been achieved in the first of the 'Window' raids.

On the afternoon of 30th July there was a somewhat relieved air conference at the Air Ministry in Berlin. Field-Marshal Milch declared: 'With his freelance night-fighting and a quickly organized technique, Major Herrmann has achieved the impossible in a very few operations, and things are rather more favourable with our night-fighting than was the case earlier.' But he warned that the enemy was preparing to intensify his attacks still further: 'What has happened to Hamburg is worse than anything that has happened to date – even during our attacks on England. The dead alone are put at 50,000 in Hamburg, as a result of the unimaginable conflagrations.' Colonel von Lossberg outlined his plans for improving on the Herrmann plan for combating the bomber stream: 'We can start with it a week from now. It is a way of keeping the real mass [of night-fighters] with the enemy all the way from the Scheldt estuary to the Ruhr and back. I feed solitary aircraft [into the bomber stream] . . . to shadow them. These shadowing aircraft will be guided as quickly as possible to the vanguard of the invading bombers. That is where *Düppel* is to our advantage, as any *Würzburg* can soon tell you where the *Düppel* track is. Each of these shadowing aircraft will be transmitting homing signals, which will attract swarms of other night-fighters to them.' As soon as the night-fighters encountered the bomber-stream they would revert to the use of airborne radar to search out individual aircraft. When they reached the target area they could switch to Herrmann's freelance tactics, and then land once they ran out of fuel. In the meantime, Kammhuber's old system of 'boxes' should remain intact, as it was invaluable for dealing with the widely scattered returning bombers.

The Germans certainly did not deserve success with their new measures. During the conference, the night-fighter experts had asked for supplies of aluminium foil strips to test their radar tactics with. At first the reaction had been that this would take at least two weeks to manufacture, and Göring was understood to have a claim on the first deliveries for a planned retaliatory blow against England. But during the afternoon Major Rüppel, one of the leading night-fighter ground controllers, was ordered to collect some supplies which had been developed by Werneuchen experimental station several months before. As Rüppel had been present at the earlier meeting, when Kammhuber had declared that jamming by 'Window' was impossible, his surprise at learning that stocks had been manufactured and held all the time was understandable.

On the evening of 2nd August, Bomber Command launched its fourth and last attack on Hamburg, but on this occasion the weather intervened. As the bombers arrived over northern Germany, they found turbulent storm clouds towering high above them and two aircraft were struck by lightning. As the brightly coloured target markers went down, they were swallowed up in the murk below, and the attack was widely dispersed. It was as though the Gods of War had cried 'Enough'. Kehrl reported that the 'detonation of exploding bombs, the peals of thunder and the crackling of the flames and ceaseless downpour of rain formed a veritable inferno'.

The number of British civilians killed in German attacks on Britain was some fifty-one thousand; in one week the slate had been wiped almost clean.

On the morning of the last great attack on Hamburg, 3rd August 1943, Field-Marshal Milch again called all the fighter experts for a conference in Berlin. Some permanent means, Milch stressed, had to be found of defeating the growing tide of British radio countermeasures: 'I am beginning to think that we are sitting out on a limb,' he said. 'And the British are sawing that limb off . . .'

Before proceeding to an examination of the aftermath of GOMORRAH, it is worth taking stock of the full achievement of the 'Window' tactic. During the first six raids in which the aluminium foil had been used – there had been two on the Ruhr as well as those on Hamburg – Bomber Command aircraft had flown 4,074 sorties; of these a very high percentage, 83 per cent, had attacked their targets, and there had been an overall loss of 124 aircraft, or barely 3·1 per cent. Early in August, General Martini reported: 'Since 25th July, the enemy has combined with his raids into Reich territory the dropping of "Hamburg Bodies" ["Window"] primarily at night but in isolated cases in daylight too. The technical success of this action must be assessed as absolute . . . By these means the enemy has delivered the long-awaited blow at our decimetric radar sets both on the ground and in the air.' In its use, 'Window' had exceeded all British expectations: as Dr Cockburn later said, they had performed a lot of trials with the tactic and knew it to be a very powerful method indeed. 'What we could not allow for were the cumulative effects which went right through the system as it lost confidence in itself.'

The Official Historians have referred to the belated introduction of 'Window' as a 'sad story'. Their view was that this countermeasure could have been employed much earlier, and they imply that it would have had much the same impact on Germans as it did in July 1943, but with a much greater overall saving in aircraft. This view is not supported by the evidence. Clearly Mrs Curran's leaflet-shaped tinfoil would not have brought the German defences to a standstill. Quite apart from its unsatisfactory 'reflecting' characteristics, sufficient could have been carried for use only in the target area. The radar stations in the main 'Kammhuber line' would not have been affected. It was not until June 1942 that Dr R. V. Jones had unravelled the full significance of the Giant *Würzburg* sites within the enemy fighter-control system, and again it was not until December of that year that it had become clear that the German night-fighter radar could fortuitously be jammed by the same size of 'Window' strips. Until this had been established, there appeared to be little doubt, as was argued by Lord

Cherwell at the time, that the British defensive system was far more vulnerable than that of the Germans.

Could RAF Bomber Command have achieved destruction similar to that in Hamburg in July 1943 had the countermeasure been used earlier? Again, this is a question of degree. The truth is that the period from 1st August 1942 to 31st July 1943 saw an enormous increase in the striking power of Bomber Command, and an improvement in its aircraft.[5] During this same period, its ability to place bombs accurately was greatly enhanced by the introduction of *H2S* and 'Oboe' at the beginning of 1943; it can also be shown that it was not until the early summer of 1943 that *H2S* training of the bomber crews was really adequate. The month of May, 1943, would seem to have been the earliest that Bomber Command could have achieved the level of destruction achieved in Hamburg. Moreover, as we are about to see, the new German night-fighter tactics were to circumvent the effects of 'Window' much sooner than had been expected.

General Kammhuber's opinion stated after the war was that the timing of the introduction of 'Window' had been exactly right. Had it been used earlier, the German radio industry might have been able to produce a whole family of new radar devices able to operate in the face of it. But by July 1943 they no longer had sufficient industrial capacity to spare – there were indeed already new pressures on the electronics industry, competing with the demands of radar. The new V-2 rocket venture was to prove the most

5. The number of bombers involved in the heaviest attack in each month during this period was as follows:

1942			1943		
	August	307		January	201
	September	476		February	466
	October	289		March	457
	November	239		April	577
	December	272		May	826
				June	783
				July	787

In mid-1942 most of the bombers were twin-engined, whilst in mid-1943 most were four-engined. The latter carried twice the bomb-load.

severe.[6] The RAF's long wait before finally introducing the countermeasure had ensured that it was not used until Bomber Command was really ready. The parallel between the introduction of the tank and the first use of 'Window' had been avoided.

By the beginning of August 1943 the British jamming attack had reduced the German night-fighter force to a state of near impotence. There were still other measures in the employ of Bomber Command, primary among which were the 'Mandrel' jammers of the *Freya* early-warning radar system and the 'Tinsel' jammers for obliterating the German ground-to-air communications. Both these devices had lost some of their initial effectiveness. An increasing number of bombers was carrying the 'Monica' tail-warning radar and the 'Boozer' radar warning receiver, and with growing effect the British bombers were relying on radar bombing aids. This was the Gordian knot the Germans had to cut.

As if this were not enough, the RAF was introducing still further tactics with which to oppress the German night-fighter force. The idea of sending out British fighter aircraft to seek out and destroy their German counterparts was not a new one; but the difficulty of making contact with the enemy in a night sky full of British bombers had

6. In this connection, a January 1943 letter from the head of the Peenemünde rocket establishment's Development Works to Heinrich Himmler's chief of staff shows the methods that were used to promote the V-2 project. Appealing for Himmler to intercede with Hitler, the writer explained: 'The V-2 rocket contains a series of electrical installations, but placing contracts for these in the electronics industry is causing difficulties as the free capacity for these is very meagre. It is vitally necessary for a Führer Decree to rank the V-2 programme higher in the priority list than the radar programme.' The writer continued by suggesting, 'This can be justified by drawing attention to the V-2's offensive-weapon character, in contrast to the purely defensive nature of the radar programme.' Adolf Hitler signed the required Führer Decree on the afternoon after Bomber Command first used 'Window', 25th July 1943.

always proven too formidable. Now that the German *Lichtenstein* night-fighter radar had fallen into British hands, however, British Telecommunications Research Establishment built a special receiving device to enable night-fighter crews to home on to the set's radiations. Code-named 'Serrate', the new device showed the *Lichtenstein*'s signals on a small cathode-ray tube: the display looked rather like a herring's backbone. In operation, the British operator would observe which side of the 'backbone' had the more numerous and bigger 'bones', and direct the pilot to turn his aircraft in that direction. When the 'bones' on both sides of the trace were equal in size and number, he knew that the target was dead ahead.

No. 141 Squadron, a night-fighter unit commanded by Wing-Commander J. R. Braham, was selected to use the 'Serrate' homer; the device was installed in the squadron's

The 'Serrate' Homer Picture

The extra length of the 'bones' on the right indicates that the German night-fighter is to the right of the attacking aircraft. This picture shows the method of *directional* indication; a second tube, turned through 90 degrees, indicated the targets relative *elevation* – if the 'bones' of the second tube were of equal length, the target would be flying at approximately the same height as the attacker.

Beaufighters, and during June 'Serrate' operations over
enemy territory began. In the weeks that had followed, No.
141 Squadron had claimed twenty-three German night-
fighters, nine of which fell to the guns of Braham himself.
One of his contemporaries later said: 'It seemed to us that
the Luftwaffe would wait for Bob Braham to get airborne,
and then come up and beg to be shot down!

All this was unknown to the German air force com-
manders, as they laid their plans to salvage what they
could from the old Kammhuber system of night air defence. Major-General Adolf Galland, the Inspector of
Fighters who had attended all the crucial Air Ministry
meetings between Milch, Kammhuber and Herrmann in the
previous few days, later summed up the feeling throughout
the air force: 'Never before and never again did I witness
such determination and agreement in the circle of those
responsible for the leadership of the German air force. It
was as though under the immediate impact of the Hamburg catastrophe everyone had put aside either personal
or departmental ambitions. There was no conflict between
General Staff and war industry, no rivalry between bombers
and fighters: only the one common will to do everything
in this critical hour for the defence of the Reich, and to
leave nothing undone to prevent a second national misfor-
tune of this dimension.'

On 31st July, Colonel-General Hubert Weise, the overall
commander of the Reich's air defences, advised all his
formations that 'the present enormous difficulties of defence
against the heavy night attacks caused by the jamming of
radar demand extraordinary measures everywhere. All
crews must understand clearly that success can come only
through the most self-sacrificing operations.' He decreed
that Major Herrmann's new fighter wing was to commence
operations over the actual target city, and that where there
was no other possibility for them to operate, all the other
night-fighter units were to be thrown into the fray over the
target area as well: they would use the tactics van Lossburg

had suggested, code-named 'Tame Boar'.[7] The fighter divisions were to inform the Air Zone and flak commands immediately their formations took off, and the latter commands were to keep the fighter divisions informed of the bombers' probable target and all developments in the night's operations. All night-fighter squadrons were to be scrambled at times calculated to bring them over the probable target simultaneously with the enemy bomber formations.

At the target itself, all means were to be used to enable the fighters to find the bombers by visual means: if there was a thin layer of cloud, a new tactic would be adopted, code-named *'Mattscheibe'* – ground-glass screen; the masses of searchlights would play evenly on the bottom of the thin cloud layer, so that the bombers could be made out in silhouette by the night-fighters circling overhead. Once the fires started at the target below, the effect would be further enhanced; radio countermeasures could not shield the bombers from tactics like these. These new 'Wild Boar' tactics meant a period of readjustment for the German night-fighter crews. Under Kammhuber's system they had nearly always operated within a few miles of their own airfields, and always with the same radar ground-control station. Everything had been very gentlemanly then. Now all that had been changed: night-fighters would journey the length and breadth of Germany hunting their quarries down, their pilots landing only when their fuel ran low.

The nerve centre of the new air defence organization shifted from the belts of ground-control radar sites to the battle-headquarters of the fighter divisions themselves. The First Fighter Division covered the east of Germany from headquarters near Berlin; the Second the north, from Stade near Hamburg; the Third the north-western approaches, from Arnhem-Deelen in Holland; the Fourth the western approaches, from Metz in France; and the Seventh Fighter Division covered southern Germany from Schleissheim near Munich. Under Kammhuber's system these headquarters

7. *'Zahme Sau.'*

had served just as 'clearing-houses' for information on the progress of the battle. Now the main flow of information was reversed: the radar sites passed news of the raids on to divisional headquarters, and not vice versa. The air situation map in the divisional operations room replaced the old Seeburg Table. One fighter control officer now ordered whole squadrons of aircraft to scramble and assemble at the radio beacons thought to be in the bombers' path. Then a running commentary was broadcast to the fighter squadrons, ordering them *en masse* to the latest positions derived for the bomber stream.

The German air force first made use of the new tactics *en masse* on the night of 17th-18th August 1943, when Bomber Command attacked the V-2 rocket establishment at Peenemünde in full moonlight. It was ideal 'Wild Boar' weather, but because of the Command's skilful tactics it brought a very near fiasco for the Germans. Fifty-five of Major Hajo Herrmann's single-seater day-fighters, and 158 night-fighter aircraft were scrambled, but because of a misinterpretation of the nature of a purely diversionary attack by eight Mosquitoes on Berlin, the chief operations officer of the First Fighter Division ordered the night-fighters to assemble over Berlin. Soon there were more than two hundred circling over the city, waiting for Bomber Command to arrive. Below them, the gunners loosed off round after round of ammunition into the wheeling aircraft, before the agonized eyes of Field-Marshal Milch who could clearly see the night-fighters firing the prescribed cartridge recognition signals; the guns kept firing, because they thought the engine noises above them were from hostile aircraft; and the fighters stayed because they thought the gunners must have information on bombers nearby. When the first target indicators went down at Peenemünde, the fighter pilots 100 miles to the south could clearly see them and they realized they had been tricked. In spite of orders to remain over Berlin, those few with any fuel remaining headed towards the glow as did those who had not yet reached Berlin. They arrived in time to catch the final wave of attackers, and shot down forty-one of them: this was a clear indication of the potential of the new system, for this 7 per cent loss was higher than those being inflicted before the introduction of 'Window'.

The Peenemünde operation brought further victories for Wing-Commander Braham's fighter squadron, No. 141 Squadron, using the 'Serrate' equipment to home on the German fighters' airborne radar. That night, five Messerschmitt 110's of *NJG 1*'s fourth squadron had been directed towards a gaggle of aircraft which seemed to have strayed south of their bomber stream. But the 'gaggle of bombers' turned out to be Braham's Beaufighters, and as the Messerschmitts moved in for the kill they received an unexpectedly hot reception. Braham turned and followed one Messerschmitt, piloted by Flight-Sergeant George Kraft, and closed in to a range of 300 yards: 'Fire was opened with a two-second burst from all guns and strikes were seen all over the enemy aircraft.' The enemy's port engine began to burn; Braham fired another two-second burst, and the Messerschmitt caught fire and dived into the sea, where it lay burning on the water. Kraft and his radar operator, Corporal Dunger, had managed to bale out into the sea; Dunger was rescued two hours later, but Kraft's body was washed ashore some weeks later on the coast of Denmark.

Almost immediately, Braham had sighted a second Messerschmitt 110, which had been chasing them. Braham curved round on to his pursuer's tail and put a one-second burst of cannon and machine-gun fire into the enemy at a range of only fifty yards. The enemy aircraft appeared to blow up, and he had to turn hard to avoid ramming it. His windscreen was flecked with oil from the exploding wreckage. Braham could only say, 'God, that was close.' Almost at once, he saw that one of the Germans had baled out and his parachute was opening. The stricken aircraft was diving vertically into the sea, on fire. In his autobiography, Braham afterwards wrote:

Perhaps it was the narrow escape from the collision

that angered me, or maybe it was because I was exhausted. I called to Jacko [the radar operator] on the intercom.: 'One of the bastards must have been blown clear. I'm going to finish him off.' I had turned towards the parachute when Jacko said: 'Bob, let the poor blighter alone.' This brought me to my senses and I felt ashamed at what I had intended to do.[8]

The Messerschmitt's gunner, Corporal Gaa, had warned his pilot that the Beaufighter was approaching, and Sergeant Vinke had pulled the aircraft into a tight turn, but too late. The exploding cannon shells tore the control column out of Vinke's hands, and both Gaa and the *Lichtenstein* radar operator were wounded; Gaa's life-jacket and rubber dinghy were cut to ribbons. Both the gunner and the radar operator baled out of the stricken Messerschmitt but neither was ever seen again. It might almost have been more merciful if Braham had opened fire. The pilot, Vinke, stayed with the falling Messerschmitt down to 9,000 feet and then he baled out. After eighteen hours in the sea, he was rescued by seaplane.

Another of No. 141 Squadron's crews accounted for a third Messerschmitt of the German squadron; it crashed on land and both the crew were killed. The fourth, piloted by Lieutenant Schnaufer, was fired upon by 'friendly' AA gunners, but managed to reach its base, and the fifth had to return early with engine trouble. It had not been a good night for *NJG 1*'s fourth squadron.

It was characteristic that these 'Serrate' actions had taken place near the sea, for the limited range of the Beaufighter prevented it from operating beyond the Ruhr or Heligoland; the new tactics of 'freelance' night-fighting introduced after the big Hamburg raids meant that the German fighter aircraft tended to operate outside the range of the 'Serrate' operations and No. 141 Squadron's success rate dropped sharply as autumn approached.

*

8. J. R. Braham: *Scramble.*

What had failed during the defence of Peenemünde finally succeeded triumphantly six nights later, during the first of a series of heavy raids on Berlin which were to bring near-disaster for RAF Bomber Command. Field-Marshal Milch had called several conferences to find out what had gone wrong on the 17th, and the blame was generally put on the flak organization. Now Göring himself had issued the most stringent orders, limiting the flak's ceiling of fire over Berlin, so that Herrmann's system could operate to maximum effect. On 23rd August, Sir Arthur Harris dispatched 727 bombers to Berlin, intent on repeating on a smaller scale the tactics which had wrought havoc in Hamburg one month before. Zero-hour was set at a quarter to midnight, but RAF wireless monitors in England listening to the running-commentary on the bomber stream's progress, broadcast by the German ground-control to all the night-fighters, heard the enemy indicate as early at 10.38 p.m. that the night's target was believed to be Berlin, and at 11.04 p.m. all German fighter aircraft were ordered to proceed to the capital.

What followed was one of those operations when the bomber-navigators seemed to be entering the phrase 'kite down' in their logs almost every minute. The bombers which survived the operation brought back reports of nearly eighty fighter interceptions and 31 actual attacking-runs by fighters on them – 23 of which were within one hundred miles of the target and 15 over the target itself. This was a most unpleasant innovation for the bomber crews, who expected to be spared this second hazard when over the flak areas. The raid was remarkable for the crews, in that they hardly noticed the flak over Berlin. At their post-raid de-briefings, many reported that the Germans had 'put up scores of fighters' and that there were 'about twenty belts of searchlights inside the capital and around it, co-operating with the fighters'. The result was a catastrophe, but not for Berlin: Bomber Command mourned its highest loss to date. Fifty-six bombers did not return, of which at least thirty-three had fallen to the German fighter-defences, no fewer than twenty of these over Berlin itself.[9]

Twice more Harris sent RAF Bomber Command to Berlin, and twice more the new German tactics cost the force many aircraft. On the night of the 31st, Herrmann had arranged for a German bomber unit to fly high over the RAF bomber stream and light it up with parachute flares for the first time, and the night-fighters were able to follow the bomber stream's own route markers placed by Mosquitoes along the attacking route. RAF wireless monitors intercepted wireless traffic indicating that aircraft were 'being drawn from areas as widely separated as Grove in north Denmark, and Juvincourt and Dijon' in France. Forty-seven bombers failed to return. On the night of 3rd September, a final attack was mounted using Lancaster bombers alone; even so, twenty more bombers failed to return.

The three raids on Berlin had cost 123 aircraft and crews, and had caused considerable damage in the western part of the city. From these three costly operations, Bomber Command drew the lesson that it was important to conceal the identity of their target as long as possible, by means of evasive routing and by feint attacks designed to seduce the night-fighters from the main target area. The new tactics forced the German air force to look to the efficiency of its reporting service; more accurate means had to be found of keeping track of the mass of the bombers, and of distinguishing the RAF's true target from the spoofs.

On the clear summer nights of 1943, Major Herrmann's target interception tactics proved highly successful. Indeed, the new freelance methods forced upon the Germans by 'Window' seemed more effective than the close-controlled tactics previously employed. Some of the cross-country flights made by the German night-fighters were quite re-

9. Mr Wynford Vaughn-Thomas, the BBC commentator, flew in one of the attacking bombers that night and made a recording of his impressions (BBC Sound Archives, Nr 6234). He sounds clearly alarmed as all around he sees aircraft blowing up, and the chaos of Pathfinder-flares, flak and tracer shells from fighter aircraft.

markable. There were cases when fighters based in northern Denmark intercepted raiders as far south as Stuttgart. The German air force's tactics were characterized by their great flexibility, and by the controller's skill in bringing the largest possible opposition to bear. The early successes attained by Herrmann's buccaneering pilots had far-ranging consequences, however, and these were not foreseen at the time: instead of concentrating adequate research effort on ridding the highly-organized German radar system of the effects of 'Window' jamming, the new night-fighter organization devoted its efforts to a series of dazzling but temporary expedients, allowing the radar-control system to fall into disuse, while all the time the British were entrenching their position still more. General Kammhuber was relieved of his command, and replaced by General Josef Schmid, commanding a newly-established First Fighter Corps. Schmid set about reorganizing Kammhuber's famous defensive system: to extend the radar tracking system over southern and eastern Germany, he closed down some of the stations in the original Line, and redeployed their equipment elsewhere. One of the first radar stations to be dismantled for this purpose was station *6B*, the site at Nieuerkerken in Belgium which had been so useful to Dr R. V. Jones during the great radar hunt in 1942.

In an effort to regain the initiative, Bomber Command introduced new radio countermeasures. One weakness of the new freelance tactics employed by the Germans was their dependence upon the fighters' getting accurate and timely information on the position of the enemy bomber stream. If the German fighter-controllers could be prevented from passing this information, the system would be set to naught. Renewed British jamming effort was therefore applied to the German ground-to-air communications channels.

The commentary on the bombers' movements was now transmitted by high-powered stations working on high frequencies of between three and six megacycles per second, and very high frequencies between 38 and 42 megacycles; for the latter range a jammer was under development, and the former frequencies were within the cover of the crude 'Tinsel' noise jammers, which transmitted engine-noises on the fighter frequencies. To jam the new German high-powered HF transmitters, Bomber Command now introduced a system known as 'Special Tinsel'. The night-fighter commentary could be received at monitoring stations in England; as soon as the German frequency had been measured, this was passed to the bombers, and their wireless operators tuned their 'Tinsel' jammers to the frequency in use. Together, they transmitted enough noise to blot out the night-fighter commentary.

'Special Tinsel' had its brief moment of glory on the night of 30th August, 1943, during the attack on München Gladbach. Within thirteen minutes of the first signals being received in England, the jamming was well concentrated on the main broadcast frequency; seven minutes later, the Germans changed this frequency and all transmissions ceased for fifteen minutes. When the commentary resumed on the new frequency, this too was jammed. That night, Bomber Command lost twenty-five of the 660 aircraft in the attacking force; the 3·8 per cent loss was a welcome improvement.

Between June and December 1943 the strength of the German night-fighter force rose from 554 twin-engined aircraft to 611, and this small rise in quantity was accompanied by a much greater one in quality as the older Dornier 217's, Messerschmitt 110's and Junkers 88's gave way to improved versions of the Messerschmitt 110 and Junkers 88 and to the new Heinkel 219. As we shall shortly see, the new aircraft were equipped with a growing range of radar and homing devices, which enabled them to intercept night bombers with increasing facility. As autumn turned to winter, there was also a potent variation on the idea of concentrating the mass of the fighters over the bombers' target: the German ground controllers began to direct fighter units into the bomber stream while the latter was *en route* to the target, with the intention of setting up long-running fights which might last for over a hundred miles. The new Y-control technique was sometimes used as an adjunct to the new method, to increase the effectiveness of the 'Tame Boar'.

In the meantime, the Germans made what progress they could with the modification of their *Würzburg* fighter-control and gun-laying radars to enable them to operate effectively in face of 'Window' jamming. They called this process 'de-lousing'. The first of the anti-'Window' modifications, code-named *Würzlaus*, emerged just one week after the RAF started dropping the aluminium foil, during August 1943. The device was demonstrated to General Martini and Colonel von Lossberg at the Werneuchen experimental station during the first week in August: *Würzlaus* employed the well-known Doppler effect whereby a radar pulse, bouncing off an object moving either towards or away from a radar set, returned with a measurable difference in frequency from the frequency at which it was transmitted. The Germans realized that when 'Window' foil was dropped from an aircraft, it quickly lost its forward speed, so the problem could be resolved by differentiating between the virtually stationary 'Window' cloud and the fast-moving aircraft. But there was an obvious snag: the *Würzlaus* device would function only where the aircraft was moving towards or away from the radar at a relative speed of more than about thirteen miles an hour. Those flying a course tangential to the set had no net speed towards or away from it, and were once again lost in the 'Window' clouds. Nor was that all, for if there was a wind speed exceeding thirteen miles per hour, the 'Window' might also appear in the same way as moving aircraft targets. Under certain conditions it was possible for *Würzlaus* to show a true aircraft echo as a 'Window' cloud, and vice-versa. The device was again demonstrated to Colonel-General Weise and General Martini early in September, and this time the results were reported as 'good'.

A second anti-'Window' modification of the radar system was devised, to overcome the snags of the other. The new modification, code-named *Nürnberg*, employed the principle that the radar pulse bouncing off an aircraft continually varied in strength because of the 'modulating' effect of the fast-rotating propellers. A man listening to the returning radar pulses through a pair of earphones would hear a distinctive rustling noise; when the set was lined up on a 'Window' cloud, this effect was not present. This second modification, which involved the addition of an extra unit to the radar to enable the pulses to be heard as well as seen, was not popular in service, as it was difficult to use properly. Only the most skilful operators were able to get reasonable results with it. On Field-Marshal Milch's instructions, Dr Hans Plendl was charged with bringing the *Nürnberg* system into service; he was to handle the project from the laboratory stages right through to the front line, and within the next three months he was able to modify 1,500 *Würzburg* sets in this way. Martini reported that the modification brought important relief to the radar-controlled searchlight units in their predicament. On 11th September, Plendl was able to report plans for combining the two anti-'Window' systems.

On the same day as Plendl made his report, the Werneuchen experimental station was able to fly a *Naxos-Z* set in an aircraft for the first time. It proved capable of picking up *H2S* radar transmissions at a range of about ten miles. Field-Marshal Milch was promised that five *Naxos-Z* sets would be manufactured each month: 'We would have enormous results, if we could shoot down the Pathfinders before the rest!' At this time the Pathfinders were the only crews equipped with *H2S* radar.

The Germans were still at some pains to find out exactly what the British crews could see on their *H2S* screen. The captured 'Rotterdam' set had made its first flight on 22nd June in a German bomber, but the cities showed up only indistinctly. The German Intelligence organization questioned hundreds of RAF prisoners to find out whether the set was used for navigation or for actual target finding, but by the beginning of September 1943 it was still an open question. Colonel Schwenke reported in Berlin that the monitoring service's *Korfu* receivers provided ample proof that the RAF's *H2S* sets were switched on throughout the whole operation, from take-off onwards. Field-Marshal Milch asked: 'Have we worked out from the PoW interrogations what the enemy thinks he can see?' Colonel Schwenke told him that it was difficult to answer this, as the prisoners had never brought pictures with them: 'So far our reports indicate only that they see very much more than we do.' Milch rejoined: 'They can very justifiably be lying, on the other hand. Any decent man would tell us a pack of lies. But it would be interesting to hear from somebody who is not a decent fellow, to find out what he can see when he's flying over Berlin.' Colonel Schwenke told him that while the PoW's agreed that it was possible to identify towns, some spoke of difficulties with the equipment, or complained that it was difficult to get a clear picture. When the *H2S* was in good order they could pick up towns at distances up to twenty-five miles away.

Field-Marshal Milch asked: 'How does the enemy differentiate between towns and the surrounding countryside?' 'They put it like this,' answered Schwenke. 'If I switch on at a distance of 50 miles, the town rises on the horizon as a bright patch about the size of a sixpenny-piece ["a tenpfennig piece"]. There was just such a report about Nuremberg.' All round the bright patch of Nuremberg, the country would be darker. General Heyne further explained: 'The town moves in like a bright moon. As the picture is always orientated with north at the top, the pilot [sic] has an excellent chance to recognize the position of the town in the picture – the "sixpenny-piece". He knows exactly where the town is. Since he knows his course, and has his target markers, he can quickly determine on the "sixpenny-piece" his target for attack.' Milch said that the best countermeasures were the night-fighters themselves; the trouble was that the RAF could bomb in bad weather using *H2S*, when the night-fighters could not operate. Someone tactlessly summed up: 'The British have tried this as a solution to the old problem of blind bombing.' 'They have succeeded!' Milch retorted. 'And they are not bombing "blind" – they can see!'

The fact remained that the RAF bomber crews were switching their powerful *H2S* radar sets on as soon as they left the English coast, even as far north as the Humber estuary and Grimsby; and the German monitoring services were swift to exploit the situation. Their Signals Command had a favourite axiom: 'All radio traffic is high treason';[10] the axiom was now applied to the enemy's radar transmissions. The Telefunken company adapted the *Würzburg* radar aerial dishes to accommodate a *Naxos* receiver for the British radar's 9-centimetre transmissions. This *'Naxburg'* receiver possessed the dual advantages that it did not radiate itself, and it had very high precision in determining the source of *H2S* transmissions. The new set was first used on 22nd September, and registered individual aircraft flying 100–160 miles away. As this first set, erected near Essen, was so successful, the local Air Zone Command ordered in October that several should be manufactured and a special organization was set up to track the RAF bomber formations across western Germany. Old *Würzburg* aerial dishes were collected, and *Naxos* receivers built into them. The dishes were mounted with their operators' cabins on a rotating base, the whole unit being highly mobile and simple to operate. Very shortly, an uninterrupted watch on *H2S* aircraft was being maintained. Two sets sited on the 2,000-foot Michelberg mountain near Münstereifel were even able to pick up the *H2S* aircraft 250–300 miles away.

In addition, a company of the German Air Force's signals regiment had found a means of triggering the 'identification, friend-or-foe' devices carried by RAF bombers, and was busy tracking the bomber formations by this means as they invaded German territory.[11]

As autumn turned to winter, the Germans had plumbed the depths of the aftermath of the RAF's introduction of 'Window', and they were overcoming their difficulties faster than their enemies had hoped. They were developing new radar equipment to counter the jamming, and homing

10. *Aller Funkverkehr ist Landesverrat.*

11. In the autumn of 1943, when British Intelligence began to mount a watch on two companies of this Signals Regiment in connection with the suspected launchings of V-weapon missiles from Peenemünde, this use of the IFF sets became known. One of the regiment's companies was indeed found to be tracking missiles fired across the Baltic; but the other was tracking RAF bomber formations by triggering their IFF sets.

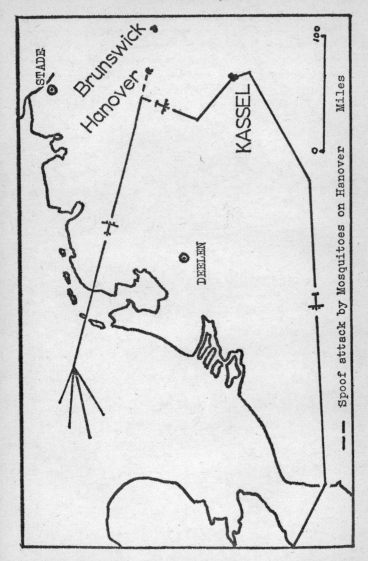

Attack on Kassel. 3rd/4th October 1943

receivers for their night-fighters to single out and destroy the *H2S*-carrying bombers. The increasing weight of electronic equipment carried in the bombers was in itself providing the German defence organization with means to follow and engage the enemy formations. Now the Germans were to recover, and defeat Bomber Command for the first time in battle.

8

The Climax

> It is very humiliating to see how the enemy is leading us
> by the nose in air warfare. Every month he introduces some
> new method which it takes weeks and sometimes months for
> us to catch up with. Once handicapped by the enemy in any
> phase of warfare, it is exceedingly difficult to catch up with
> him. We have to pay very dearly for what we failed to do
> in air warfare hitherto. But that was to be expected.
> – DOCTOR GOEBBELS' diary entry, 7th November 1943

General Josef Schmid, the German night-fighter force's
commander, could follow the movements of the bomber
stream from take-off until the target was reached, and then
back again, thanks to the British practice of leaving the
H2S radar set switched on throughout the operation. From
October onwards, the new airborne radar code-named *Lichtenstein SN-2* was also in operation; the *SN-2* had a maximum range of from $2\frac{1}{2}$ to 4 miles, but its importance was
primarily that it was not affected by the 'Window' then
in use. Moreover, the German tactics of hurling the night-
fighters *en masse* against the bomber stream were causing
the bombers heavy losses.

RAF Bomber Command countered by tactics of deception in depth. The tactics were typified by the operation
against Kassel on the night of 3rd October 1943 : the force
of 540 heavy bombers and nine Mosquitoes flew in a straight
line from the coast towards Hanover. Two German fighter-
controllers gave Hanover as the target, and ordered their
fighters to assemble there. But at a point some thirty miles
to the west of the city, the bomber stream turned to the
south, while the Mosquitoes alone carried on. These latter

dropped green target-markers and a few scattered bombs on Hanover, forcing the German night-fighters to stay there in case a full-scale attack developed. When the main force failed to appear, and the markers started to go out, the controllers ordered their fighters to head for Brunswick, and it was not until eighteen minutes past 9 p.m. – seven minutes after the first yellow markers burst on Kassel – that the German 'running commentary' mentioned that city. The first fighters arrived there only just in time to catch the final waves of the attack. Twenty-four bombers, 4·4 per cent of the force dispatched, failed to return.

The German night-fighter force should have been able to do better with its new tactics against the British bombers; on this particular occasion, the co-operation between the anti-aircraft guns and the night-fighters had been poor, and the Germans had not made the best use of their radar tracking stations. The reasons for the failure were fully dealt with during a Berlin conference on 5th October, with Field-Marshal Milch in the chair. Colonel von Lossberg, who had himself been piloting a fighter aircraft two nights before, described the kind of thing that had happened:

'The flak reports: "Heavy formation of enemy aircraft approaching", and then the order is given "Open fire!" and they let loose their gunfire to all manner of altitudes. As a result one of the squadron's aircraft based at Venlo had its engine shot out over Bonn, and another was damaged over radio-beacon *"Otto"* – both by flak, and this just to one squadron. I myself came up against flak bursts and was forced to weave.

'During the attack on Kassel, the flak was so intense at every altitude that most of the night-fighters would not close right in to over the target. They waited outside in the dark, and in consequence had little chance . . . Orders had been received to fly round these flak zones, but how is that to be done in cloud? The failure during the Kassel raid is, however, basically due to other reasons: the enemy formation flew in by an unusual route to Kassel, and they made it very difficult for us because they sent in a feint attack [on Hanover]. They flew in on a course which suggested an attack on Hanover and then, about forty miles west of Hanover, the enemy formation turned south and flew from the east over the Harz mountains towards Kassel.'

Milch asked the obvious question – why the night-fighters had in that case not pursued them. A representative of the Air Zone Command told him that there was no proper radar in the area. Milch was not satisfied: 'That can't be the reason. Even if there was not complete radar cover, it must have been possible to establish the course of the enemy formation. I cannot understand why nobody noticed the enemy turn off forty miles west of Hanover. I am making no attacks on anyone personally, nor on any organization. I shall only say this: the system is not in order. I don't care who is to blame, it's all the same to me. But what does make me feel bitter is the fact that we cannot maintain control during our night operations, although hundreds of thousands of people are involved.'

As for the night-fighters having been fired on by the anti-aircraft guns, he said it was obvious that the mutual co-operation system was not functioning: 'I am convinced that the commands are being given, but unfortunately they are not always carried out. It is all wrong. Ultimately, defeat or victory hangs on this question. Unfortunately this does not seem to be appreciated everywhere.'

By now, RAF Bomber Command was introducing further radio countermeasures to bedevil the German night-fighter control system. From the spring of 1943 there had been a growing need for a jamming transmitter to cover the frequencies from 38 to 42 megacycles – the range of the VHF radios fitted to the German night-fighters. Dr Cockburn and his jamming team at the Telecommunications Research Establishment at Malvern had now produced such a device –'Airborne Cigar'. For its operation, 'ABC' required a full-time operator and there was thus provision for an extra crewman in the Lancasters of the squadron to which the device was issued, No. 101 Squadron. The ABC 'whiskers', seven-foot long spa-aerials – two on the upper fuselage and one under the nose – gave the bombers the appearance of pre-historic monsters, but surprisingly they had little effect on the Lancasters' performance. ABC aircraft accompanied all major bombing raids from October onwards. One of the jammer operators recalls:

In the Lancaster we sat in darkness, cut off from the rest of the crew. The nearest other crewman was the mid-upper gunner – his boots were at my eye-level, about four feet further down the aircraft.

We had a monitor receiver with a three-inch-diameter cathode-ray tube. This represented the frequencies from 38 to 42 megacycles as a horizontal line, and any signals we picked up were shown as vertical blips which grew out of the base line. When we saw a signal we would tune our receiver in to it and listen. If it was a German message we would switch on one of our three transmitters and tune it to the receiver signal so as to obliterate it. The immediate result was generally a small movement of the German signal up or down my tube, and it was just a question of following it. If the jamming got too bad, the German would go off the air altogether. Later he would appear on a different frequency, but it was a matter of seconds to cover the new blip.

Often the German controller would get rather angry. I sometimes felt a little sorry for him, trying to get through while I was stopping him.

It was not long before the Germans had overcome the worst effects of the earlier 'Special Tinsel' jamming introduced to cover the normal German fighter frequencies. The Germans installed more very high-powered transmitters and broadcast the 'running commentary' simultaneously on several frequencies. This had the effect of reducing the jamming power available for any one channel. But the RAF had yet another card ready to play: signals in this part of the frequency spectrum, from three to six megacycles per second, do not travel only in straight lines; by bouncing off the layer of ionized gas which surrounds the earth they will carry for several hundred miles, bending to the curvature of the earth as they do so. It was thus possible for high-powered ground transmitters in England to give jamming support to RAF bombers penetrating deep into German territory. Several transmitters at Rugby and Leafield, belonging to the Cable and Wireless Company, the GPO and the BBC, were modified to cover the frequencies used by the German night-fighter force. Special aerials were erected to beam the signals eastwards, towards Germany. The transmitters' frequencies could be remote-controlled through a landline link by the RAF monitoring station at Kingsdown in Kent.

Originally, the intention had been to use the new device to transmit jamming noise, but speech transmissions were also possible. If the night-fighters relied on instructions from the ground to find the bombers, could not fake instructions be transmitted to them to send them astray? It seemed an effective and economical method of reducing the bombers' losses, but during the ensuing discussions it became clear that such tactics might prove to be a double-edged sword: while it was possible that some of the German aircrew might be confused by fake orders regarding the target, it was unlikely that the German fighter-controllers, working in comfort, with good information on the actual track of the attacking force, could be deceived in any way. Indeed, the contradictions of any British 'ghost' controller might serve as a valuable guide to the actual intentions of the attackers. If the target was Leipzig and the German controller had ordered his fighters to assemble there, while still privately entertaining some doubts as to whether that was the target, all his doubts would disappear if the 'ghost' voice then tried to divert the fighters to Brunswick or some other imaginary target. The danger was that information could be given away by implication. Though many of the fighter crews might act on the false order, the more experienced crews would be able to follow the true picture. And with his doubts removed the German controller could endeavour to concentrate his forces on the now known target. It might be argued that the possibility of a double bluff did exist, but not enough could ever be known of the information available to the German fighter-controllers for such subtlety to pay off.

As a result, the 'ghost' controllers at Kingsdown had strict orders to make no mention of any real or imaginary routing of friendly aircraft, nor to a target whether genuine or otherwise. But within these limits there was plenty of scope for ingenuity. Thus, during the second of the three earlier air raids on Berlin the Kingsdown monitors had noted that when night-fighters were informed of fog at their airfields, they tended to break off their operations early. The 'ghost' voice could broadcast false warnings of fog, and endeavour to bring the night-fighters home earlier than they had intended; other similar tactics could also be employed.

The new tactic, code-named 'Corona', went into operation on the night of 22nd October 1943 for the first time. The Germans first guessed that a big attack was being mounted for the night when 'J' beams were switched on in England; at a quarter to seven that evening, bomber squadrons were detected by long-range radar leaving England, and at five past seven several hundred aircraft were reported crossing the coast between the Katwijk and Scheldt estuaries. The bomber stream's course was plotted towards the south-east as far as the Rhine, and isolated *H2S* bearings confirmed its track. A small number of Mosquitoes continued to the south-east, but the Germans could see that they were only Mosquitoes, because of their high speed. Just south of Bonn, German reconnaissance aircraft saw the bomber stream release a string of flares to mark their turning point and from that moment on the RAF's diversionary attack on Frankfurt had failed. When the bomber stream had first been tracked, it had been heading for Frankfurt, and the night-fighters of three Fighter Divisions – 179 aircraft – had been ordered to a position north-west of Frankfurt, ready to pounce on the bombers over the target. Now that the course-change had been seen, the fighter-controller ordered all fighters to Kassel – and Kassel was the real target that night. The night-fighter squadrons hurried north with a following wind, and wrought havoc on the RAF bombers, despite all attempts of the 'Corona' controller to draw the

German night-fighters away. Some north German squadrons had actually stumbled across the head of bomber stream while flying to the earlier Frankfurt position.

The German fighter-controller recognized that fake orders were being fed into the system by the British, because of the faulty pronunciation. He warned the night-fighters: 'Don't be led astray by the enemy' and 'In the name of General Schmid I order all aircraft to Kassel'. Becoming increasingly angry, the German controller actually swore into his microphone, causing the 'ghost' controller in England to remark: 'The Englishman is now swearing!' The German controller shouted: 'It is not the Englishman who is swearing, it is me!'

Because of the German reconnaissance aircraft's sighting success in detecting the force's change of course towards Kassel, the Frankfurt 'spoof' attack was a failure, and Bomber Command suffered heavily: forty-two bombers (6·9 per cent) were lost, of which 25 were from the first two waves. But even so, Kassel had been very effectively destroyed in a raid which cost several thousand lives and destroyed many important armament factories.

Gradually the most effective form for 'Corona' was evolved. For successful night-fighting it was important that the crews should be able to operate their equipment calmly with the minimum of interference, so anything which raised the level of annoyance was an asset to Bomber Command. The 'ghost' controller would call up German air and ground radio stations, giving long test transmissions: 'One-two-three-four-five. Five-four-three-two-one. Monday-Tuesday-Wednesday-Thursday-Friday-Saturday-Sunday. Testing-testing-testing.' Anyone who has had to listen to such drivel on a loudspeaker system will testify to the annoyance it can cause. On the night-fighter crews waiting for instructions at the assembly beacons, the effect was even greater: one by one, Germany's cities were being razed to the ground, and each night they were listening to somebody broadcasting test-transmissions. By requesting test-transmissions from the Germans themselves, it was possible for the 'ghost' to tie up frequencies for several minutes at a time. After a few weeks, the Germans ceased to respond to these requests, so the 'ghosts' would read out pieces from Goethe, turgid bits of German philosophy, and even play gramophone records of Hitler's speeches. All pretence of imitating the German fighter-controllers was given up.

On 3rd November 1943, Air Chief-Marshal Sir Arthur Harris minuted Mr Churchill:

> We can wreck Berlin from end to end if the USAAF will come in on it. It may cost us 400–500 aircraft. It will cost Germany the war.

This was the kind of promise that Mr Churchill could not resist, and the RAF was authorized to launch its third great 'Battle', the Battle of Berlin. The American bomber force was still smarting from the severe losses it had suffered while attempting to hit German targets in daylight, and they decided not to 'come in on it'. Harris decided to go it alone.

Though the nights were longer than they had been in the summer, the Reich capital was still a formidable target for the bomber force. It involved a round trip exceeding eleven hundred miles, so that if the bombers took the most direct possible route with no changes of course to confuse the defences, they would still have to fly over 360 miles of enemy territory, at least an hour and a half's flying-time. The first attack was launched with 444 heavy bombers on 18th November, and only nine failed to return; this rate was promisingly low. The Command returned to the city three more times during November, and four times in the following month, and the losses stayed surprisingly moderate – possibly in consequence of the bad weather over Germany at the time. By 28th November, the Germans had one aircraft equipped with the *Naxos-Z* device for homing on *H2S* transmissions, and although it was currently proving impossible to distinguish between individual bombers with this device, the aircraft was still of use for shadowing the bomber stream and homing in the rest of the night-fighter force: it was capable of picking up *H2S* aircraft in the bomber stream from over 60 miles away.

The Germans had given great thought to the problem of protecting Berlin from British bombers, especially when they attacked by radar alone. In a report on Berlin's vulnerability to *H2S* radar, distributed to Göring, Milch, Martini, Speer and others, Plendl wrote that while the erection of radar decoy sites might marginally hamper the bombers' radar navigation to the capital, it would never prevent it altogether. A jamming transmitter had been designed, codenamed *Roderich*, but it needed far higher power to be of effect against *H2S*: 'The only effective measure remaining to us is the active engagement of the aircraft equipped with *Rotterdam* [*H2S*] by "homing" receivers. But these will enjoy success only so long as the enemy continues to leave his *Rotterdam* continually switched on.' Throughout the summer, Milch had discussed with his experts the possibility of 'camouflaging' Berlin's numerous lakes and waterways from the penetrating stare of the *H2S* carried by the bombers. The conclusion finally reached was that the project would require thousands of tons of steel and other materials, and would never really succeed. Recourse was made to more orthodox forms of deceit: anti-aircraft guns fired specially modified anti-aircraft shells to explode at different altitudes into the blaze of coloured flares characteristic of British Pathfinder marking; but the German explosives experts had considerable difficulty in copying accurately the colours of the British markers – their red was in particular far too dull. Even so, these decoy markers did deceive inexperienced crews on occasion. The effort put into the German decoy fire sites was even more impressive: no fewer than fifteen such sites ringed Berlin. The largest was fifteen miles to the north-west of the city, an exceedingly elaborate affair measuring nine miles in diameter – a full-sized replica of Berlin in plywood and cardboard, complete with a fake Tempelhof airfield. Simulated bomb explosions, houses on fire, searchlights and anti-aircraft fire combined

with the 'spoof' Pathfinder markings to give these decoy
sites a strong element of realism.

On the night that the Battle of Berlin had opened, one of
the 'Airborne Cigar' Lancasters had been shot down, and
its equipment fell into German hands. The unusual jamming equipment was removed from the wreckage, and sent
to Telefunken for examination. On 30th November, Colonel
Schwenke reported on the find to Field-Marshal Milch,
during a conference on the RAF's jamming offensive:

> I have some interesting foreign Intelligence, concern-
> ing a number of new things that have come to light. Three
> transmitters have been found in the Lancaster aircraft –
> in fact it was flying not with the normal seven-man but
> with an eight-man crew, one of whom was an additional
> radio operator. The set is called T.3160[1] and is apparently
> there for the jamming of our VHF radiotelephone traffic.

Schwenke recalled how the enemy had begun jamming
the Germans' *FuG10* night-fighter radio-telephone late in
1942, by means of a normal transmitter coupled to a microphone in the bomber's engine-bay; this was the RAF's
'Tinsel' device, of course. The Germans had obviated the
effects of this jamming by going over to a new VHF transmitter and receiver, the *FuG16*, and this had worked until
the new ABC (Airborne Cigar) jamming had been introduced. Of the captured ABC transmitter, Schwenke said: 'It
is a special VHF transmitter which cannot really be used
for any other purpose as it is not employed in radio traffic
at all itself. At present I cannot say how far this jamming
can be defeated by appropriate countermeasures or frequency alteration!'

Milch asked the obvious question: 'How far can these
enemy jamming transmitters be used for our fighters to
home on to?' Schwenke replied that obviously any transmitter used by any aircraft could be homed on to, once its
frequency had been established. Milch astutely pointed out

1. The RAF designation of 'Airborne Cigar'.

that the jamming transmitters would be tuned to the Germans' own frequencies. The trouble was, he was told, that the RAF jamming aircraft switched their transmitters on only for short periods, so homing on them was impossible: 'Quite apart from this, the enemy now tries to upset radio communication by cutting in on the traffic himself.' Milch asked: 'What are we doing about this?' He was told: 'For a week now we have been using girls to broadcast the radio-telephone instructions; that means the enemy would have to take girls into action with them!'

Evidently the Germans believed that the 'ghost' voice was airborne, like 'Cigar'. The RAF was already countering this latest German move, by using German-speaking WAAFs to mimic the German girls.

While Bomber Command was about to come as close to defeat as it had ever come since Sir Arthur Harris' appointment as Commander-in-Chief, it was still possible for the RAF to have very much the upper hand in the radio counter-measures offensive. The British had taken the initiative, and in Milch's graphic phrase the Germans were 'trotting after' them. A bitter quarrel developed between Milch and Martini during December on the RAF's great jamming war. Milch protested: 'When I read your own comments on this, Martini, I must say I find them bursting with such pessimism that I am forced to ask, is it really justified? The Reichsmarschall declares: "Apparently we can't jam anything and the British can jam everything!" And all you say is: "Yes that's about it!" ' Martini retorted that the reason for his pessimism was that they had so far been unable to fly radio-reconnaissance missions over England, and moreover that German industry no longer had the manpower or capacity for new electronic developments.

Milch proclaimed: 'The Reichsmarschall [Göring] looks at it differently. He says quite openly: "On our side we've only got woolgatherers! The British can do things much better than us!" For a while that might have been true, with a set like their *Rotterdam* [*H2S*] ... Only yesterday I was reading a very interesting report on an interrogation of a

PoW on the use of the radar set during their latest attacks:
they attacked Berlin, and did this by picking up the lakes
near Brandenburg first, then the lakes near Potsdam, and so
on.' In any case, Milch admitted, the British were a seafaring
nation and born navigators. The German air force on the
other hand had its roots in the army, and you could not call
stopping at every crossroad and looking at the signposts
'navigation'.

On 16th December, the RAF attacked Berlin again. Again
the 'J' beams gave the Germans early warning of a heavy
raid, and the bomber stream was clearly tracked by its own
H2S radar transmissions all the way; again the small diversionary attacks by Mosquitoes were recognized as such
because of the Mosquitoes' high speed, and the fact that
there were no *H2S* transmissions from them. But the weather
was unfavourable for night-fighting, and the German fighter
Corps suffered other impediments as well:

> Corps VHF radio jammed by bell sounds, radio-
> telephone traffic hardly possible, jamming of Corps HF
> radio by quotations from Hitler's speeches, Corps alterna-
> tive frequency and Division frequencies strongly jammed.

Nor was that all, for there had been 'Very sudden jamming of the "Anne-Marie" forces broadcasting station by
continuous sound from a strong enemy jamming station.'
This latter move was not pure spite, for the RAF listening
service had correctly deduced from the unbalanced musical
selections being offered from this Stuttgart broadcasting
station that the programmes were being used as a crude
means of indicating to the night-fighters the position of the
bomber stream: waltzes meant that the bombers were in
the Munich area, jazz meant Berlin, church-music Münster,
Rhenish music Cologne, and so on. At the departure of the
attackers, the station regularly broadcast the march 'Old
Comrades' and then resumed normal transmissions. As soon
as this crude – and it must be admitted desperate – device
had been recognized by British Intelligence, a special high-
powered transmitter, 'Dartboard', was brought into use to

jam the music; it was this that the Germans had heard.

As a last resort, the Germans took to broadcasting instructions to their night-fighter force in Morse code. Morse signals are the most difficult of all to jam, but the move availed the Germans little because the majority of the night-fighter radio operators had not had proper training in using the Morse code. In any case, the RAF reacted to the German move by setting up transmitters in England, appropriately dubbed 'Drumsticks', capable of broadcasting strings of completely meaningless Morse signals on the German frequencies.

Only by broadcasting their orders simultaneously on a number of channels could the German fighter controllers avoid the worst effects of this jamming barrage. This in turn meant that the frustrated aircrews had their work cut out reaching the spectrum for unjammed frequencies, and even when these were found there was no guarantee that they would remain that way for long. By the beginning of 1944 the combined offensive mounted by the 'Tinsel', 'Corona', 'Airborne Cigar', 'Drumstick' and 'Dartboard' jammers had brought chaos to German night-fighter communications.

The high-precision bombing attacks on the Ruhr still mystified the Germans, as they had not managed to bring down a Mosquito with an intact 'Oboe' set.[2] They hoped that they would soon have the answer to the high-altitude Mosquitoes, as by injecting nitrous-oxide (which they called *GM1*) into the engine of the Junkers 88-R at the critical moment they could give it the requisite height performance and an excess speed of about 25 miles per hour over the Mosquito's. At present, all the Germans knew was that a radio monitoring station near Essen had observed quite by chance that the release of Pathfinder flares during a certain attack on Cologne had occurred just as the Morse letters *5T* (. . . . . —) were picked up on the wavelength employed by the high-

2. 'Oboe' was the RAF's ground-controlled precision radar bombing aid. See page 139.

altitude Mosquitoes; and that early in November, 21 repeated Mosquito bombing runs on the Bochum Union steelworks had been observed. At a conference between the
works' civil defence officers, local flak and radio monitoring
experts, the flak officers reported that they had clearly seen
the aircraft following a 'gently curving track, centring on
the southern coast of England'. Again the individual bomb-
release points had coincided with the Morse letters *5T*.
This seemed to clear up one mystery, about which the
Führer had himself expressed an opinion: he had asked
the Intelligence services to find out whether the fact that
the targets for these precision attacks were usually steelworks
meant that some kind of infra-red homing system was used.
Milch referred a few days later to 'an extraordinarily accurate attack on the August-Thyssen furnaces'. Only five
bombs had been dropped and all had hit. Reichsminister
Speer said: 'It certainly is remarkable that for these last
two months in an increasing number of isolated nuisance-
raids, 6, 8 or 10 bombs have been dropped, usually through
heavy cloud cover, of which 80 to 90 per cent have hit blast
furnaces or power stations.'

The SS and Intelligence services were called in on the
Führer's instructions to establish whether in fact enemy
agents might have planted radio beacons near the vital
targets. The commission of investigation discovered from
a small number of unexploded bombs among those dropped
by the Mosquitoes that they were not radio-controlled in
any way: that meant that the Mosquitoes themselves must
be controlled from England. Milch was told: 'The British
have special Mosquito crews expert in radar technique, who
have been under training for long periods, and they drop
their salvoes from altitudes of 25,000 to 30,000 feet blindly
into the works. They have two devices on board for this,
and with these they make two distance-measurements; and
it is through these receivers that the instruction is radioed
for the bombs to be released. It is up to us to jam this system
at once.' Jamming alone would not suffice for long, as the
German Intelligence services had deduced that the British
already had alternative frequencies ready.

On General Weise's instructions, a jamming transmitter code-named '*Karl*' was readied to jam the 'Oboe' frequency already in use, but for technical reasons the jamming proved more difficult than expected. In the meantime, a *Freya* set was employed south of Duisburg in the Ruhr to determine the exact starting point of the Mosquito's bombing run. In each case, it was found to be eight minutes' flying time from the final target. This just gave time for the target's sirens to be sounded. Jamming transmitters were set up on either side of the 'Oboe' tracks, but they had little effect. During a heavy all-Mosquito attack on the Krupp works on 12th December, 90 per cent of the bombs were direct hits despite the jamming. Still no 'Oboe' set had been captured, so the Germans could only continue jamming and hope for the best.

It was not until the end of December that the Germans found a partial answer to the RAF blind-bombing system. During an 'Oboe' attack on Rheinhausen, the Germans ordered the jamming frequency to be 'swept' [*variiert*] up and down the spread of frequencies used by the British; because of a hitch, the first Mosquito attacked accurately as before, but then the swept jamming was put into effect. The second Mosquito began its bombing-run correctly, but then the distant signals from England became unsteady and vanished altogether before the release signal *5T* could be given. The radio monitoring station at Essen reported: 'Swept jamming successful. The aircraft became uncertain and broke off prematurely, mostly without bombing.'

On 7th January, an 'Oboe' Mosquito was brought down near Kleve, and its equipment fell into German hands. Now the Germans could see that the 'Oboe' system was a twin-channel system, and they knew how best to jam the control signals. Within three days plans had been laid for a network of about eighty jamming transmitters to transmit 'noise' over the frequency range 220 to 250 megacycles per second. This would give the British least hint that the 'noise' was in fact deliberate jamming of 'Oboe'. The result was evident within the next few days, as the accuracy of the 'Oboe' bombing operations fell from 90 per cent hits to less than 25 per cent.

At the end of the month, however, the Germans were given the first remarkable hint that the British knew what was up. A German radio monitoring station following the transmissions on the 'Oboe' frequency picked up a Parthian shot tapped out in Morse code by one of the ground transmitter operators in England on the actual 'Oboe' frequency:

Achtung, Achtung! You are a *Schweinhund*!

The German Intelligence service deduced from this crude linguistic effort that the British were fully aware of the German jamming and it 'indubitably means that conversion [of "Oboe"] to the centimetric wavelength is to be expected. And this cannot be jammed'. On 30th January a *Naxburg* operator definitely picked up 9-centimetre signals being transmitted by a high-flying Mosquito back to England. However, this evidence was discounted, the more so when the radio monitoring service continued to pick up the old wavelength 'Oboe' signals; the possibility that the latter might be being transmitted just for camouflage did not occur to the Germans for several months, and 'Oboe' operations continued on the new wavelength unhindered.

For the time being, as Reichsmarschall Göring had complained, the truth was that the Germans could not jam anything and the British could jam everything. During December, Telefunken's Dr Rottgardt was forced to admit that the technique of ridding the radar system of the effects of 'Window' was still inadequate: 'At present our technology has no answer to "Window" – and certainly none that could be employed in an aircraft. In fact we are not working on one, and I don't see any way in which one could.' The best solution to the problem would have been a centimetric airborne radar on the lines of the American *SCR720*. But German technique on such short wavelengths was still insufficiently far advanced: the only solution was the *SN-2* night-fighter radar which worked on a frequency as yet un-jammed by 'Window', 90 megacycles. Its maximum range of four miles, and its wide-angle of cover, were valuable improvements on the standard *Lichtenstein*, but the high minimum range of 400 yards had made the set virtually unacceptable until it was recognized as the only set not affected by 'Window'. The poor minimum range received the most urgent attention, but it was early 1944 before a solution was found. In the meantime the new sets were taken straight off the Telefunken production line and installed in the night-fighters; to enable the crews to follow bombers within the minimum range, a simplified version of the older *Lichtenstein* was also fitted. The weighty installation sprout-ed a forest of aerials on the aircraft's nose, but there was no workable alternative.

The new year brought a new overlord for radar research in Germany, as Dr Hans Plendl retired from his post as Plenipotentiary for High-Frequency Research; Professor Abraham Esau, who had just been removed from his office as Plenipotentiary for Nuclear Physics, was appointed to replace him. During his one-year period in office, Plendl had certainly achieved much: he had put radar research on a much sounder footing than it had been in 1942, and he had laid the foundations of a centimetric radar system; he had worked tirelessly on the problems of anti-'Window' devices, and the 'invisible U-boat'. But he knew that he had been fighting a losing battle. In a letter to Himmler on 7th January 1944, he warned that what the Western Powers had brought about with their tenfold greater capacity, their vastly superior financial resources and raw materials, and their industries untouched by bombing, the Germans could only hope to equal by enthusiasm and self-sacrifice: 'In my view, this is the only magic charm that can break the spell in the long run.'

One of Esau's first proposals was that a nation-wide com-petition should be held – in secret – for the best method of defeating 'Window'. Competitors were invited in an official German Air Force circular to 'give a method and means whereby either the jamming can be obviated or a distinction can be drawn between the jamming "targets" and the true target when large clouds of "Window" are used'. The solutions offered were to be 'technically feasible' and clear, and both the problem and the solution were to be kept a close secret to those outside the groups involved. Göring himself offered six tax-free prizes of between about £1,300 and about £16,000 for the best solutions, which were to reach the Air Ministry in Berlin not later than 1st April 1944. In the months that followed, Esau's department for high-frequency research investigated no fewer than twenty-five different anti-'Window' schemes. The schemes met with varying degrees of success, but none seems to have been perfected in time for the competition's closing date as none of the prizes was in fact awarded.[3]

Even without an answer to 'Window' the Bomber Command losses were beginning to rise alarmingly as the new year opened. Sir Arthur Harris was committing his entire force in an attempt to destroy Germany's main cities, and in particular Berlin; the revitalized German defences were determined to stop him so doing. The night-fighter crews were flying to defend their own homes and loved ones with no less bravery and determination than had their counterparts in the RAF during the Battle of Britain in 1940. It was no longer on just isolated occasions that the British bombers suffered heavily.

The first of the great and costly battles was on the night of 21st January, when 648 bombers set out to bomb Magdeburg, near Berlin. Fifty-five aircraft failed to return: not since the disastrous attack on Berlin itself five months before had the RAF suffered so heavily. The bomber force had fought back hard, but the German losses for the night amounted to only seven machines. One of the German fighter-aircraft was a Junkers 88 piloted by the famous

3. Even now, thirty years later, 'Window' remains a useful jamming method.

Major Prince zu Sayn Wittgenstein, holder of the Knight's
Cross with Oak Leaves and the highest score of victories to
date, 83 'kills'. His radio operator, Sergeant Ostheimer,
described in his report how he took off at about 9 p.m. on
a 'Tame Boar' type operation that night:

At about 10 p.m. I picked up the first contact on my
[*SN-2*] airborne radar. I gave the pilot directions and a
little later our target was seen: it was a Lancaster. We
moved into position and opened fire, and the aircraft
immediately caught fire in the left wing. It went down at
a steep angle and started to spin. Between 10 and 10.05
p.m. the bomber crashed and went off with a violent explosion. I watched the crash.

Again we searched. At times I could see as many as
six aircraft on my radar. After some further directions,
the next target was in sight – another Lancaster. After
the first burst from us there was a small fire, and the
machine dropped back its left wing and went down on
a vertical dive. Shortly afterwards I saw it crash. It was
some time between 10.10 and 10.15 p.m. When it crashed
there were heavy detonations, most probably it was the
bomb-load.

After a short interval we again saw a Lancaster. After
a long burst of fire the bomber caught fire and went down.
I saw it crash some time between 10.25 p.m. and 10.30
p.m. The exact time is not known. Immediately afterwards we saw yet another four-engined bomber. We were
in the middle of the 'bomber-stream'. After one pass this
bomber went down in flames; at about 10.40 p.m. I saw
the crash.

Obviously, the new type of radar was not being affected
by any of the bombers' countermeasures. Within a few
minutes, Ostheimer again had a target 'blip' on his radar
and after a few course alterations yet another Lancaster
came into sight. After one attack, the bomber began to burn;
but the fire appeared to go out, and as Major Wittgenstein
moved into position for another attack a series of explosions

rocked his Junkers 88 and its left wing began to burn. The Junkers 88 began to fall out of the sky. 'As I saw this the canopy above my head flew off and I heard on the intercom a shout of "Raus!" – Get out! I tore off my oxygen mask and helmet and was then thrown out of the aircraft.' About fifteen minutes later, Ostheimer landed by parachute, but the Prince was trapped in his crashing Junkers 88 and his body was found in the wreckage. Only the night before, he and Ostheimer had gone in so close below a Halifax to make certain of hitting it with their upward-firing cannon that the plunging bomber had crashed right into them; they had been lucky to escape with their lives. This time they had not been so fortunate: they had almost certainly been attacked by another of the bombers in the stream, for none of the five long-range fighters of No. 141 Squadron made any claims that night. The Prince seems to have fallen victim to a surprise attack from below – the form of attack he himself had used to such effect.

Not far from the scene of the Prince's crash lay the wreckage of a Heinkel 219 also destroyed that night. Its pilot was also dead – Captain Manfred Meurer, the third highest scorer among the night-fighter pilots with 65 kills. Meurer died when the bomber he had just shot down from below crashed into him; this was clearly a major hazard of such close combat.

By early February, the radar equipment of the night-fighter force was showing a considerable improvement, and the ground organization was still able to rely on its *Mammut*, *Wassermann* and later *Freya* sets. Moreover, the night-fighters were making increasing use of the *H2S* tracking devices. Twenty-eight *Naxos-Z* sets had been delivered of thirty-three that were on order, and these would enable the night-fighters into which they were installed to home on individual *H2S* aircraft and shoot them down. Most promising of all was the equipment of the night-fighters with the new *SN-2* radar: by 1st February, two hundred had been fitted and several hundred more were on their way.

No *SN-2* had yet fallen into British hands, and it was still unjammed.

The German defences took an increasingly heavy toll of the attacking bomber force. On 28th January, 43 bombers had failed to return from 683 attacking Berlin, and the month that followed was even worse. On 15th February, 42 were lost out of 891 attacking Berlin and four days later the Command lost 78 out of 823 attacking Leipzig, a success attributed by the Germans entirely to the new *SN-2* pursuit night-fighting procedure.

On most occasions, the German fighter controllers saw through the British attempts to divert the fighters away from the raiders, but there was one spectacular exception on the night of 20th February. That night a force of about two hundred aircraft from operational training units flew out over the North Sea almost up to Heligoland. Nearly the whole of the German night-fighter force assembled over southern Denmark in anticipation of yet another attack on Berlin. The German controller only realized his mistake when the operational training units' aircraft turned round and went home, and a much larger force of bombers was reported nearing the Rhine – the 598 bombers of the RAF's main attacking force, which had flown in over France without arousing undue interest. When he realized what was happening, the controller ordered all his fighters to make for Strasbourg. He could have saved his breath, as the distance involved was 400 miles, one and a half hour's flying time away. Those fighters which did arrive with any fuel in their tanks did so long after the bombers had departed. The sharp attack on Stuttgart cost the RAF just nine bombers. This kind of deception could not succeed often. The Germans learned to delay their interceptions a while until the position was quite clear; this was something the night-fighters could well afford, since the bombers spent up to two hours over Germany during the long penetration attacks.

During the follow-up attack on Stuttgart, on 15th March, thirty-six bombers were lost out of 863.

Bomber Command next penetrated deeply into Germany, to Berlin, on the night of 24th March. The Germans concentrated their forces around the capital in good time and the raiding force paid heavily. Seventy-two bombers out of 811 dispatched were shot down. Even the successful attack on Essen two days later, when only nine out of 705 failed to return, was overshadowed by this.

RAF Bomber Command visited Germany once again on 30th March. This was to be Sir Arthur Harris' final attempt to smash a major German city before the operational control of his force was turned over to General Eisenhower to prepare the way for the forthcoming invasion of Europe. The night's target was Nuremberg, and 795 bombers set out to attack it, following an unusual course that was to take them in an almost dead straight line for hundreds of miles across enemy territory, with one last turning-point just before Nuremberg. Even before the leading aircraft had crossed the coast of England, the German listening service had determined very accurately the course of the bomber stream by following its *H2S* transmissions. The Mosquito spoof attacks on Cologne, Frankfurt and Kassel and several other minor targets were all complete failures, as the Germans identified the aircraft concerned quite clearly as Mosquitoes: they were flying without *H2S* radar. The German ground controllers exploited this early Intelligence to the full, and made every effort to 'switch in' the German night-fighter squadrons into the bomber stream as far to the west as possible. The Third Fighter Division's controller ordered all twin-engined night-fighters to assemble over radio-beacon *'Ida'* to the south-east of Achen, and these squadrons were fed into the bomber stream.

By midnight, when the RAF bomber stream began to thunder over the *'Ida'* radio beacon, a German night-fighter force of 246 aircraft was airborne, and nothing could prevent the slaughter that followed. Once again, the German radio communications were beset by every kind of jamming – bell sounds, 'quotations from Hitler's speeches', noise and shrieks. Again Stuttgart's *Anne Marie* transmitters were blotted out, but with the night-fighters in the air nothing could prevent them from seeing the bomber force: the winds were much higher than had been forecast in England, and the bomber stream had begun to lose its cohesion; even before its first major turning point it was moving on a front forty miles wide, and due to the unusual meteorological conditions worse was to follow. Each minute, the petrol burned in one aero-engine produced one gallon of water as steam; normally the steam dispersed, but on this cold night it condensed and the long white condensation trails of vapour suspended in the sky chased remorselessly behind each bomber as it crossed the Rhine to the east. It was a clear night, and the glow from the half moon gave the vapour trails a phosphorescent quality.

The Third Fighter Division joined battle over its radio beacon, and a running fight began which was to last for the next two hundred miles. A bomber squadron of *KG7* had also arrived, a special 'illuminating' unit, and these aircraft unloaded strings of parachute flares high over the RAF bomber stream; the flares could be seen by fighters all over Germany and they converged on them like moths round a flame. The Second Fighter Division arrived from northern Germany, and the First from the Berlin area, moving westwards on a collision course. The Seventh Fighter Division came up from southern Germany, via radio-beacon *'Otto'*. It was an ideal night for these 'Tame Boar' tactics, and the night-fighters wrought fearful execution on the bomber force. One of a Pathfinder aircraft's crew recalls his pilot exclaiming at one stage, as their bomber was still on its easterly leg: 'Better put your parachutes on, chaps. I have just seen the forty-second one go down.' At the bombers' altitude there was a fifty-mile-per-hour tail wind which took many of the bombers far to the east of their intended track, particularly during the final south-easterly leg of the approach to the target.

The Germans consequently had great difficulty in discovering which target the bombers were making for. Not until eight minutes to one, two minutes before the bombing was due to start, did the 'running commentary' mention Nuremberg for the first time. By then the first bombs had

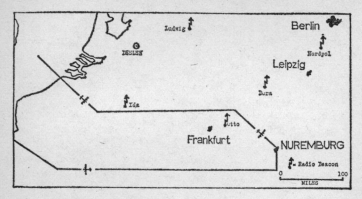

The Attack on Nuremberg

in fact already fallen, for some aircraft had reached the target ahead of schedule because of the strong tail wind.

The single-seater 'Wild Boar' units were all too far from Nuremberg to go into action, as things turned out. Three squadrons were orbiting over '*Otto*' in anticipation of an attack on Frankfurt and four squadrons were waiting over beacon '*Nordpol*' to cover the twin threat to Berlin and Leipzig.[4] Thus only the twin-engined night-fighters went into battle; twenty-one squadrons – about 200 fighter aircraft – saw action.

Few bombs actually fell in Nuremberg that night, and they caused little damage. So widely scattered was the bomber force as it withdrew that the German night-fighters lost it almost completely, and the surviving bombers returned virtually unmolested. For Bomber Command the cost of the Nuremberg raid – the high point in Germany's pursuit

4. Since the war there have been suggestions that the Germans had some foreknowledge that the night's target was to be Nuremberg and had arranged their defences accordingly. Whatever the truth of the former suggestion, the dispositions of the 'Wild Boar' units make it clear that the latter was incorrect. Moreover, most of the 'kills' scored by the twin-engined night-fighters were before the bombers reached their target; had foreknowledge been the cause of the disaster, the bulk of the 'kills' would have been at the target or after it.

night-fighting phase – was very high indeed: ninety-four Lancasters and Halifaxes failed to return, nearly all of which had been shot down along the route between radio beacon '*Ida*' and the target. This portion of the track had been clearly marked by a trail of aircraft blazing on the ground.

Thus ended the Battle of Berlin: the Reich capital had not been destroyed, although Bomber Command had lost more than twice the number of aircraft spoken of by Sir Arthur Harris in his note to the Prime Minister five months before. Germany was still very much in the war, but Bomber Command was not: during the thirty-five major attacks between 18th November 1943 and 31st March 1944, the Command had lost 1,047 aircraft and a further 1,682 were damaged. Whatever its outcome, the Nuremberg attack would have seen the end of the battle, for the heavy bombers were now needed to knock out targets in France and the Low Countries in preparation for the forthcoming invasion of Normandy. But even had Harris been allowed to continue his deep-penetration attacks on Germany, it is difficult to believe that his Command could have continued to accept such punishing losses.

In the night skies over Berlin, Magdeburg, Leipzig and Nuremberg the German night-fighters had avenged their humiliation over Hamburg in the summer of 1943. They had demonstrated their remarkable resilience, and had come close to making the destruction of their homeland too expensive for the RAF. This round in the war had definitely been won by the Luftwaffe.

9

In Support of Invasion

Confusion now hath made his masterpiece.   – *Macbeth*

Whatever the outcome, the long-awaited invasion of northern Europe in June 1944 was bound to be a major turning-point of the war. If it should fail, the Allied losses would almost certainly be such that Hitler need not fear for his western territories for some considerable time; but if it should succeed, then the end of the Third Reich would be in sight. Those were the issues, and they were important enough to warrant the preparation and reservation of the most elaborate radio-countermeasures that the British could devise.

The introduction of some revolutionary new counter-measures to paralyse the defences, as 'Window' had nearly a year before, would have been ideal; but the truth was that there was no new tactic likely to cause confusion like that evoked by the little aluminium strips over Hamburg in the summer of 1943.

This was not to say that Dr Robert Cockburn and his staff at the Telecommunications Research Establishment had nothing up their sleeves, for they had been preparing for just this moment ever since the darkest days of 1940. They now believed that by the imaginative use of radio countermeasures it might be possible to divert the Germans' attention away from the real landing areas in Normandy during those critical hours while the first troops fought their way ashore, punched through the coastal defences and established a defensive perimeter. What was important was that the invasion, D-Day, was to be a once-for-all affair. The

most blatant forms of trickery – tactics which could be
expected to work only once – could be confidently prepared.

If the seaborne invasion was to achieve any sort of surprise, then the first priority was obviously to be attached
to the destruction or deception of the dense network of
radar stations erected along the French coast as part of the
Germans' formidable 'West Wall'. On the northern shores
of France and Belgium there were no fewer than ninety-two
radar sites keeping watch on hostile shipping, and some of
the sites were equipped with the whole menagerie of German ground radars – *Mammut* and *Wassermann*, the Giant
and standard *Würzburgs*, the *Freya* and the naval radar,
Seetakt. If the multiplicity of these radar 'eyes' was to make
the problem of jamming more difficult than anything previously attempted, the overall problem of deception promised to be even more formidable: to keep the Germans
guessing for half an hour as to the real target of an RAF
bomber force moving at 225 miles per hour was one thing;
but to conceal a mighty invasion fleet making only ten knots,
for ten hours and more, was quite another.

A comprehensive programme of radio countermeasures
was devised. This was to fall into four parts, of which the
first two were part of a general 'softening up': the German
radar stations would have to be located, and the majority
would have to be 'taken out' by air attack. On the actual day
of the invasion, there would have to be feint 'invasion forces'
to draw the Germans' attention away from the areas being
attacked, and those radar stations remaining operational in
the real invasion area would have to be rendered useless by
jamming.

By the spring of 1944, Dr R. V. Jones' scientific Intelligence department had built up a very comprehensive picture
of the German coastal radar network, but this picture had
to be kept constantly up to date: radar sets, especially the
mobile ones like *Freya* and *Würzburg*, could move quickly
and be operational within hours of their arrival at a new
site. To assist the plotting of radar stations, Dr Cockburn's
staff produced a special ground direction-finder which could
measure the bearing of an enemy radar transmitter to an

accuracy of a quarter of one degree: three such sets, code-named 'Ping-pongs', were set up along the southern coast of England, and these provided a number of fixes on German radar stations. Once their position had been triangulated in this way, the existence of the radar stations was confirmed by photographic reconnaissance.

While Jones kept watch for new enemy radar sites, the Second Tactical Air Force began the task of 'taking out' the more prominent of those sites already known. The first of these attacks was made on 16th March, when twelve Typhoons of No. 198 Squadron set about the huge *Wassermann* early-warning set erected on the Belgian coast near Ostend: the formation crossed the coast shortly after midday at 8,000 feet as though bent on attacking inland targets; once inland, however, the four leading aircraft peeled off into a dive and streaked in at treetop height towards the towering aerial, while the other Typhoons strafed the flak emplacements surrounding the radar station; the leading four aircraft launched sixteen rockets each and scored several hits on the aerial structure before making good their escape. As they thundered into the distance, the 130-foot *Wassermann* tower was still upright, so shortly before 4 p.m. the squadron again took off to attack the tower. Several more hits were scored, but the battered tower still stood, apparently unscathed. The Achilles' Heel of the *Wassermann* tower lay in its turning mechanism, however: the aerial tower was fixed to a rotating sleeve, which turned in a fixed vertical cylinder; the rockets' blast had so distorted this sleeve that it would not turn, and the aerial could not be moved. Unfortunately for the Germans, the tower could be lowered to the ground for repairs only when it faced in a certain direction, so it now had to remain rigidly locked in the vertical position. The entire aerial structure had to be dismantled before the Germans could begin repairs, and in fact this Ostend set was still off the air when the D-Day invasion began.

The RAF soon found that the *Mammut* also had its weakness. The rear of *Mammut*'s aerial was a mass of feeder cables which had to be carefully adjusted to give the correct shape of beam; the whole set had to be checked against a series of calibration flights during which aircraft flew carefully set patterns. The feeders themselves occupied an exposed portion on the back of the aerial, and even small-calibre machine-gun fire was sufficient to cut them to ribbons, putting *Mammut* out of action until the whole tedious setting-up process had been repeated.

In the period immediately prior to the invasion, Mosquitoes, Spitfires and Typhoons of the Second Tactical Air Force flew a total of nearly two thousand sorties against German radar targets, often at a high cost in aircraft; they succeeded in putting out of action all but sixteen of the original ninety-two sites and all the narrow-beam *Mammut* and *Wassermann* sets, which were the most difficult to jam. In the invasion area proper, no radar remained in full operation when D-Day dawned.

In the meantime, Dr Cockburn and his team had put the finishing touches to the most elaborate piece of spoofery to be attempted in the jamming war – the simulation on the enemy's surviving radar screens of two full-scale 'invasion fleets'. The simplest and most reliable method would have been to use a lot of full-sized ships, but the invasion proper was stretching Allied shipping resources to the utmost and there were no ships to spare for the purpose. Cockburn worked out a method of producing a radar response which would look just like a real invasion fleet, without calling for the use of ships at all. Basically, he had to put up a radar reflector to cover a square with sides sixteen miles long – an area of 256 square miles – and get it to move across the Channel at some eight knots: the reflector would consist of a pattern of 'Window' clouds at specially calculated spacings, which would be sown by bomber aircraft flying through the square. Mrs Joan Curran, who had conducted the initial 'Window' trials in 1942, did the mathematical calculations for this extraordinary spoof.

Wing-Commander Derek Jackson, of the 'Window' Panel, devised a completely new type of 'Window' for the operation. Since the naval *Seetakt* and the *Freya* ground-radar sets covering the coast of France both worked on a much longer wavelength than the AA gunlaying *Würzburg*, the 'Window' to confuse them would have to be somewhat longer than that already in use. This was not as simple as might at first have appeared, for the longest 'Window' would now have to be nearly six foot long, which would have been quite unmanageable in an aircraft. Jackson evolved a system of folding the strips concertina-fashion into manageable lengths; after they had been released from the aircraft the weighted end ensured that the zigzag unfolded properly to the correct length.

Taking the *Seetakt* as the more difficult of the two radars to be spoofed, the ghost fleet had to be designed to be effective against that radar: if it worked on *Seetakt*, it would certainly fool the *Freya*. The naval radar's beam had a width of fifteen degrees, so at a range of ten miles the beam was $2\frac{1}{2}$ miles wide. Allowing a small margin for error, it was planned to place the 'Window' clouds within two miles of each other along the width of the ghost 'fleet', as this arrangement would produce one continuous 'blip' on the *Seetakt* screen. Again, the perception in depth of the German radar was such that it could not discriminate between targets less than 520 yards apart, so to get a continuous 'blip' in depth the 'Window' clouds had to be closer than that. The bomber aircraft sowing this ghost 'fleet' in the sky would be flying through the square at 180 miles per hour, or three miles a minute; so if the 'Window' was dropped at a rate of twelve bundles per minute this would result in one cloud being sown every 440 yards, which was sufficient for the purpose.

Altogether, the operation required eight aircraft per ghost 'fleet', divided into two waves, flying in line abreast with two miles between each aircraft, and eight miles between each wave of four bombers. To simulate the advance of the 'fleet', the two waves of aircraft would fly a series of oblong patterns, exactly maintaining this formation; each oblong measured eight miles by two. At the end of each seven-minute orbit, the formation was to move forward one mile, so the formation's actual advance in one hour was about

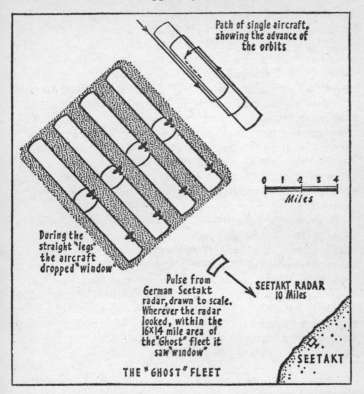

Path of single aircraft, showing the advance of the orbits

0 1 2 3 4
Miles

During the straight "legs" the aircraft dropped "window"

Pulse from German Seetakt radar, drawn to scale. Wherever the radar looked, within the 16×14 mile area of the "Ghost" fleet it saw "window"

SEETAKT RADAR 10 Miles

SEETAKT

THE "GHOST" FLEET

eight miles. On the German radar screens this would create the impression of an invasion fleet advancing at eight knots: during the long legs of the aircraft orbits, when they would be flying towards or away from the German radar, the crews were instructed to drop 'Window' at a rate of twelve bundles per minute. To add a final touch of realism, other aircraft would be orbiting nearby, operating 'Mandrel' jammers; but the positions of these latter aircraft would be chosen so that the Germans could just see through the chinks in the blanket of jamming, and discern the approach of the 'invasion fleet'.

Thus far all was theory. The spoofing force was low on the

list of priorities when aircraft came to be allocated for the
invasion, and it was not until the beginning of May 1944 that
Cockburn gained control of two bomber units, No. 218
Squadron with Stirlings, and No. 617 – the 'Dam Busters' –
Squadron with Lancasters. He visited each squadron in turn
and told them what he wanted. Then he arranged for the
crews to rehearse the complicated flight patterns. While
they were training, he arranged with Mr J. B. Supper to
move the captured *Würzburg*, *Freya* and *Seetakt* radar sets
up to Tantallon Castle on the Firth of Forth.[1] Here they
went on the air again, far from the curious stare of the
German radio monitoring service. The ghost 'fleet' spoof
was tried out against the sets and found to be completely
successful.

To prove the system finally, Cockburn had the bombers
fly a spoof against the radar site at Flamborough Head,
where there was a Type II radar – the nearest British equivalent to the Giant *Würzburg*. The operators, who had not
been briefed on what to expect, all agreed that the blip on
their screen must have come from a very large convoy
indeed – a convoy larger than they had ever seen before. Dr
Cockburn was now confident that his ghost 'fleet' would
deceive the *Seetakt*, *Freya* and the small and Giant *Würzburg*.

While the captured German radar sets were at Tantallon
Castle, a new device from Cockburn's workshop was tried
out; 'Abdullah' was a radar homer, enabling the aircraft
which carried it to pinpoint the exact location of the *Würzburg*, the smallest of the German ground radar sets. A
Typhoon was fitted with 'Abdullah', and trials began using
the captured set at Tantallon. Squadron-Leader Hartley,
flying the aircraft, found that the new device gave ranges

1. The move was not without incident. Supper's team had continued to refer to the *Würzburg* by its original German designation
FuSE62 – Funk Sender-Empfänger 62. Because of a typist's error
on the dispatch papers, however, this became 'Fuse Type 62',
quantity one. It took a lot of fast talking before the constable at
the main gate at Farnborough would accept that this title really did
refer to the 1½ tons of secret ironmongery he was allowing to escape.

exceeding fifty miles on the *Würzburg* provided that it was approached from the seaward side, and with sufficient accuracy for an immediate rocket attack to be made. The problem was that it was on the seaward side that the German radar had its maximum range, and if the defences were alerted any formations of aircraft led into the attack by 'Abdullah' aircraft could expect heavy losses. A tactic was devised, whereby the 'Abdullah' aircraft would slip in frontally and release a smoke bomb on the radar site, while the main formation of aircraft approached at low level from the rear; but even this ran into snags, for Hartley found that the 'Abdullah' had to be spot-tuned to the target's frequency before take-off, and the Germans made a habit of taking the selected target radar suddenly off the air, or switching on another radar on the same frequency at an entirely different site.

Even if the selected radar stayed on the same frequency, the German operators invariably switched off their sets as soon as they saw a high-speed 'plot' making straight for them, leaving the 'Abdullah' aircraft with no signals at a range of five or ten miles, 'a disconcerting experience', as Hartley described. The idea was never finally put into effect on a large scale. In any event it did not matter much, for the triangulated 'Ping-pong' plots and Jones' Intelligence sources, supported by photographic reconnaissance, provided all the information that was necessary for the fighter-bombers to destroy the radar sites before D-Day.

While the Tactical Air Force was waging a war of attrition against the radar sites, Bomber Command was dealing with other vital radio targets outside the fighter-bombers' capabilities. First to be attacked was the powerful radar jamming station on Mount Couple near Calais, which might have been used to cover a German counter-attack. This was the station which had jammed the British radars covering the Channel during the dramatic escape of the German warships from Brest two years before. A week before D-Day, a force of 105 Lancasters succeeded in putting down seventy heavy bombs within the radar jamming station's compound, which measured only 300 yards by 150 yards. Mount Couple had made its last mischief, and the 'Channel Dash' score was settled.

Two other large communications stations, at Au-Fevre near Cherbourg and Berneval-le-Grand near Dieppe, suffered a similar fate. Perhaps the most important target in this series of attacks was the headquarters of the German listening service in the west, at Urville-Hague near Cherbourg. It was demolished by ninety-nine heavy bombers of the RAF shortly before D-Day.

It was still light on the evening of 5th June, the eve of D-Day, when the first part of Dr Robert Cockburn's spoof programme opened: it was a refinement of the ghost 'fleet' hoax, and it very nearly did not come off. To ensure that German aircraft with their own high-discrimination radar sets would not be able to see through the spoof 'fleet' tactic, Cockburn had resuscitated a device which had not been used since 1942 – 'Moonshine'. This was a device, it will be remembered, which picked up the pulses of enemy radar sets, amplified them and sent them back like the radar echoes from a very large object indeed. Four RAF air-sea rescue launches had been supplied for the use of the Malvern jamming team, and in each of these Mr D. J. Allen-Williams had fitted one 'Moonshine' set. The four launches were to cruise underneath the falling 'Window' of the ghost 'fleet' and return the signals from any German radar sets that might be airborne in the area. The launches had arrived at Tewkesbury, the nearest they could get to Malvern, only during the last week in May, and things were eventually cut so fine that they were due to be completed and arrive at Newhaven on the very day that the invasion force was to set sail. Because of bad weather, the launches in fact arrived one day late, but as D-Day was postponed by one day anyway, in the event that did not matter.

Now, while the greatest seaborne invading force in history made final preparations to move against the enemy, the four 'Moonshine' boats rumbled noisily out of Newhaven. Each towed a float above which swung a barrage balloon almost as big as the launch itself: this was 'Filbert', a 29-foot-long naval balloon with a 9-foot diameter radar reflector built inside its envelope. 'Filbert' produced a radar echo similar to that from a 10,000-ton ship. Fourteen small naval launches accompanied the 'Moonshine' boats, each flying one 'Filbert' and trailing a float to which was attached another. Once clear of Newhaven harbour, the strange armada split into two: three 'Moonshine' launches and six of the small boats made for Cap d'Antifer, which was to be the apparent target of one of the ghost 'fleets', operation TAXABLE; and one 'Moonshine' launch and eight boats made for Boulogne, the apparent target of the second, operation GLIMMER.

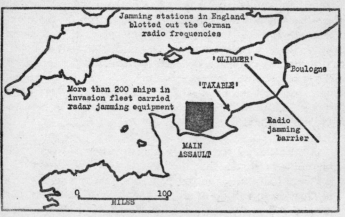

Radio Countermeasures in support of the Invasion of France

Just after midnight, the 'Moonshine' operator in one of the GLIMMER force's RAF rescue launches, No. 1249, observed signals on his cathode-ray tube from a German aircraft's radar. He tuned in his transmitter, switched it on, and the game was on. During the next two hours he logged signals from eight separate aircraft, and all but one of them, of very short duration, he 'Moonshined'. Fifty miles to the west, the 'Moonshine' operators in the TAXABLE force also picked up German radar transmissions. They too returned them with 'interest'.

Over these small boats flew the Bomber Command Lancasters and Stirlings weaving their complex patterns and sowing the 'Window' for the two ghost 'fleets' in the air: No. 218 Squadron flew the GLIMMER spoof, and No. 617 Squadron the TAXABLE one. The latter squadron's Lancasters navigated their complex tight oblong courses by means of *GEE*, while the 218 Squadron Stirlings were equipped with the new radar device, *G-H*; this was a system rather like 'Oboe' in reverse: *G-H* aircraft carried a small radar transmitter and receiver, and again the system's range was extended by a repeater transmitter, but this time the repeater was on the ground. The system's advantage over 'Oboe' was that several aircraft could operate on the same system simultaneously, which was of course vital for the ghost 'fleet' spoof flown by No. 218 Squadron on the night of 5th–6th June. It was eloquent evidence of the advances made by British radar technique by mid-1944 that a squadron of aircraft could be ordered to orbit for three or four hours over such a precisely defined, constantly moving square of the Channel's surface, keeping a regular open formation all the time. One of the No. 617 Squadron crewmen later recalled:

At the time I was rather concerned about what the Germans would do when they saw TAXABLE. We knew that we were bait and expected just about every night-fighter in creation to roll up at any moment. Our Lancaster was packed full of 'Window', from nose to tail. If we were forced down in the sea there would be little chance of our getting out before the aircraft sank.

We finished 'Windowing' at 4 a.m. by which time it was just beginning to get light. The sky seemed to be full of transport aircraft and gliders: the 'Red Berets' were going in. We hoped we had made things a little easier for them.

In the event the spoof 'fleets' were flown with no interference at all from the German night-fighters. Dr Cockburn himself was seriously exercised by the possibility that the Germans would send out an aircraft to look at the TAXABLE and GLIMMER areas, for the enemy would then see with his own eyes that there was really no invasion at all. Cockburn need not have worried, for such calculations made no allowance for the general fog of war. The very sophistication of the German reporting system was its weakness. One frightened young conscript would see what appeared to be an invasion fleet through the jamming on his radar screen, and he and other operators down the coast would report to their superiors; eventually their reports would end up as a broad arrow on the situation map at German headquarters. It would be a confirmed fact. Even if an aircraft were to fly over the area and the crew reported seeing no ships, they would not be believed.

When the two ghost 'fleets' arrived at the stop lines some ten miles off the coast of France, the small boats' crews moored their 'Filbert' floats. The naval launches then laid a smoke-screen and broadcast over loud-speakers a recording of the squeals, rattles and splashes germane to a number of ocean-going ships dropping anchor. Now their work of deception was complete, and the boats cleared out of the area.

While TAXABLE and GLIMMER were moving laboriously towards the coast of France, other mischief had been afoot: twenty-nine Stirling and Halifax bombers of Nos. 90, 138, 149 and 161 Squadrons staged a fake airborne invasion in the Caen and Cap d'Antifer areas under the code-name TITANIC. *En route* to the 'dropping zones', the bombers' crews had dropped bundles of 'Window' to increase the radar echo from the force, and in the dummy landing areas they unloaded special fireworks designed to explode in a way resembling a ground battle in progress. A few men from the Special Air Service were also dropped with orders to make a lot of noise.

Meanwhile, the mighty armada of aircraft laden with real airborne troops droned in towards their dropping zones near Normandy, and these would have presented a perfect target to German night-fighter crews: 1,069 heavily

laden transport planes, the great majority had neither armament nor the means to detect an attacking aircraft's approach. If only a few fighters could have intercepted them,
their pilots would have wrought savage destruction on the
invaders. The lumbering trains of glider aircraft would have
been even more vulnerable, for at the first sign of danger
their tugs' crews would have had to cast them adrift and
leave them to their fate.

On either flank of the transports' route, scores of Mosquito night-fighters thirsted for the chance to do battle.
But these could give no guarantee that determined German
fighter crews would not penetrate their escort lines. The
RAF took every precaution to distract the attention of the
German fighter forces on the vital night: twenty-four Lancaster bombers of No. 101 Squadron and five Flying Fortresses from No. 214 Squadron generated for the German
radar system a ghost 'bomber stream' along the line of
the River Somme. On board each aircraft the 'Window'
man tossed out his bundles of tinfoil for all that he was
worth, and while he did so the special equipment operator
working in each bomber's belly intently watched the monitor
screen of the 'Airborne Cigar' jamming transmitters. Between them the aircraft of this decoy force carried no fewer
than eighty-two ABC transmitters, enough to ensure that no
signals within the jammers' cover would be heard over
northern France. Other jamming transmitters situated in
southern England blotted out the remainder of the communications frequencies used by the German night-fighters.
From now on, if any night-fighters should make contact
with a transport plane it would be a chance meeting. There
would be no help from the ground.

The German fighter-controllers ordered all available aircraft to intercept the ghost 'bomber stream', but once the
fighters had entered the fog of jamming they could not be
contacted again. One of the No. 101 Squadron Lancasters
was shot down, but its crew was saved; and all the time the
heavy transport planes were unloading their human cargoes
over Normandy and turning back to England.

In the invasion area itself a veritable powerhouse of jamming was deafening the organs of any German radar sets that might have survived the mayhem of the previous weeks. Over two hundred of the ships escorting the fleet carried jamming transmitters, and all were switched on. There was nothing subtle about this, the final trick in the Allies' radio countermeasures repertoire. It blinded the defenders as cruelly and effectively as pepper thrown into their eyes. In the event, only one German radar station actually caught a glimpse of the approaching armada, but such was the general level of confusion that its warning went unheeded.

The first certain indication that something real was moving towards the coast of Normandy came at 2 a.m. on the morning of 6th June: observers stationed on the eastern side of the Cherbourg peninsula heard with naked ears the rumble of ships' engines. No conceivable jamming effort could have achieved more than had been done, for the invaders' approach had passed undetected until then. The Germans did mistake the GLIMMER operation for an invading force, and they opened fire on it with radar-controlled coastal guns and E-boats were ordered to the area. Operation TAXABLE appeared to draw only slight attention. It was too close to the main invasion for its effects to be observed with certainty; possibly the preceding fighter-bomber attacks in this area had been so thorough that no radar sets remained undamaged to see the approaching 'fleet'.

The rest of the story is history. Once the Normandy invaders had established a beachhead ashore, no power at Hitler's command could dislodge them. Had the Germans had the unhindered use of their powerful network of radar stations, the defenders – whose bravery was beyond question – would have reacted far more violently than they did. The Allied lives thus saved were indeed bought cheaply. Such was the level of confusion created in the German reporting system that when news of landings in the Normandy area came through it was thought that these too were feints. The defenders continued to wait for the main assault to come. It was not until the afternoon of D-Day that the Germans finally committed their armour to the battle, and by then the Allies were ashore in strength.

The final round in the jamming war had opened with the Germans being floored by the first punch. The punch would open a wound from which there was to be no recovery – a wound which RAF Bomber Command was to exploit to the full.

10

End of One War

'Confound and Destroy'   – No. 100 Group's motto

It was in the middle of July 1944 that the first complete *SN-2* radar set fell into the hands of British Intelligence, and confirmed that there was a new and efficient airborne radar in service with the German night-fighter forces. In the six months prior to April 1944, the number of contacts on 'Serrate', the homing device fitted to RAF long-range night-fighters, had fallen considerably; this in itself had been an indication that the Germans were taking the standard *Lichtenstein* radar out of service, because of its susceptibility to 'Window' jamming. The continuing heavy losses suffered by the RAF bombers *en route* to and from their targets were evidence enough that the Germans had replaced the standard *Lichtenstein* with something better. As we now know, Göring had ordered a crash programme for the production of *SN-2* radars, and the one-thousandth set had been delivered to the night-fighter force on 10th May.

British Intelligence suspected that this device, '*SN-2*', was not the only innovation, for early in the year German air force prisoners had begun to mention two other devices fitted to their night-fighters, code-named *'Flensburg'* and *'Naxos'*. Their exact nature was far from clear, so the RAF dispatched radio reconnaissance aircraft over enemy territory to listen for transmissions from these sets; the task proved more difficult than had been the case in the search for *Lichtenstein*, for *SN-2* operated in the same part of the radar frequency spectrum as the more recent *Freya* early-warning radar sets, and aircraft hunting for new signals in the right place picked up a whole family of pulses all seemingly familiar and accounted for. On top of this, *Flensburg* and *Naxos* were passive homing devices, like the RAF's 'Serrate', and emitted no radiations at all.

The first piece of hard evidence on a new German radar came from the camera-gun of an American long-range fighter which had been escorting bombers during a daylight attack. The American pilot had pounced on a Junkers 88 night-fighter and shot it down. When the camera-gun's film of the engagement was processed, it was noticed that the nose of the German aircraft was disfigured by a strange aerial array. The photograph provided Dr R. V. Jones' Intelligence office with plenty of material for speculation, but they had still neither seen the set itself nor heard its radiations.

Then, by a remarkable stroke of good fortune, the problem was solved by the Germans themselves. At 4.30 a.m. on 13th July 1944, a lone twin-engined aircraft circled the airfield at Woodbridge in Suffolk. The runway controller took it to be a Mosquito and flashed a green 'clear to land' signal to the aircraft. The plane touched down and taxied to the end of the runway, where it switched off its engines. Its crew were standing around on the apron, stretching their legs, when the crew bus arrived to pick them up; and it was in this way that an RAF flight-sergeant came to find himself confronted with three live German air crewmen. The surprise was mutual, but the British NCO hurriedly produced a Very pistol and forced the Germans to surrender. The 'Mosquito', it now transpired, was a fully-equipped Junkers 88 night-fighter. Its inexperienced pilot had inadvertently steered a reciprocal course on his compass and arrived in England without knowing it. He had been lucky to reach Woodbridge; when RAF technicians attempted to take a sample of the fuel from the aircraft's tanks they found that there was insufficient even for analysis.

The captured Junkers 88 was full of electronic equipment ominously unfamiliar to British Intelligence. It was equipped with both the new *SN-2* radar and *Flensburg*, the homer which enabled night-fighters to use the radiations from RAF bombers' 'Monica' night-fighter warning radar equipment. Most important was the discovery that *SN-2* worked on a frequency of 85 megacycles, which meant that the standard 'Window' used for the last year had had no effect on it. Fortunately, the concertina-like 'Window' developed by Wing-Commander Jackson for the invasion 'fleet' spoofs was effective against this new frequency, and within ten days Bomber Command had started using the new zigzag foil. The jamming immunity enjoyed by *SN-2* during its eight-month operational life had come to an end.

This left the *Flensburg*'s potentialities to be examined. In addition to his work on 'Window', it will be remembered that Jackson was an expert on night-fighter equipment: it fell to him to evaluate the new German homing device. He arranged for one Lancaster with its 'Monica' switched on to fly away from Farnborough on a westerly heading at 15,000 feet, while he orbited Farnborough airfield in the Junkers 88 at the same altitude. He was still picking up the Lancaster's 'Monica' signals when the bomber was over Exeter, 130 miles away. The signals were very clear, and gave a precise indication of direction; Jackson found no difficulty in using them to home right on to the bomber. To establish whether the jumble of signals from several 'Monicas' close together would confuse the *Flensburg*, Jackson repeated the trial using five borrowed Lancasters. The tuning was again so fine that homing presented no problem.

Now the danger of the British tail-warning set was clear: it was more likely to result in losing bombers than saving them. News of Jackson's trials quickly reached the ears of Air Chief-Marshal Harris. He asked Jackson: 'What would happen if there were about a hundred bombers, all using "Monica"? Wouldn't there then be a confusion of the signals?' Jackson replied that he could not answer these questions from a trial involving only five aircraft. To give a reliable answer, he would need a much larger number of bombers at his disposal. Harris agreed and said he would arrange it. At the end of August, seventy-one Lancasters flew the circular route round Cambridge, Gloucester, Hereford and back to Cambridge, all with their 'Monica' sets

switched on. Once again Jackson orbited Farnborough airfield in his Junkers 88, watching the screens of the *Flensburg*
apparatus. He picked up the first bombers at a range of
forty-five miles, and the chase was on: using *Flensburg*
alone, he directed his pilot right on to individual Lancasters
time and time again; once more the fine tuning of the German homing device made short work of the mass of signals.
Harris took characteristically decisive action after the demonstration: he ordered the immediate removal of the tail-
warning radar from all aircraft in his Command.

At the same time, the Command finally realized the full
dangers of the indiscriminate use of airborne transmitting
equipment in the bomber force. As a result, bomber crews
were forbidden to switch on their IFF identification equipment except in emergencies and the *H2S* radar was not to
be switched to transmit until the bombers were within forty
miles of enemy territory and within German ground radar
range in any case.[1] This move – which to the reader will
have seemed long overdue – made life much more difficult
for the German air force for the jamming had forced their
raid-tracking organization to lean heavily on just these
sources of Intelligence. By the time this step was taken by
the RAF, the Germans had set up a chain of *Naxburg* stations[2] along the whole front line from Schleswig-Holstein in
the north to the Black Forest in the south; one set mounted
on high ground at Feldberg in the Black Forest had succeeded again and again in tracking the RAF bomber formations by their centimetric radar transmissions all the way
from the Channel to southern Germany.

While these systems had been passive, relying on the radiations from the bombers' own radar equipment, the Germans
had by the summer of 1944 also perfected a new ground
radar set called *Jagdschloss* – and this had from the outset
been designed to work in the face of hostile jamming. It

1. Crews operating 'Mandrel', 'Tinsel', and 'Airborne Cigar'
jammers had already been under strict orders to confine their
jamming transmissions to the time they were near or over enemy
territory.
2. See page 230 above.

operated on one of four previously set frequencies within the spread of 120 to 158 megacycles per second: if the enemy jamming was very serious on one frequency, the operator could select each of the three alternatives in turn until he found one which was clear. *Jagdschloss* had a range of ninety miles, good enough for the loose control of night-fighters engaged in 'Tame Boar' operations.

Most ingenious of the German attempts to beat the jamming was the *Klein Heidelberg* device. Strictly speaking, it was not a true radar set as it did not transmit. Instead, it used the transmissions from one of the radar stations situated in England to 'illuminate' targets. The principle was a simple one: the *Klein Heidelberg* receiver was locked to the direct transmissions coming from the British radar station but in addition it picked up echoes from the aircraft in the area. The receiver was connected to an indicator tube which therefore displayed two 'blips' – one representing the distance from the British radar to the set and the second the distance travelled by the reflected echo-wave from the British radar to the receiver in Germany, via the aircraft. Since the former distance was fixed and known, it was possible for the receiver operator to calculate the second by simple arithmetic: he could plot the aircraft's position as being somewhere on an ellipse whose foci were the German and the British stations; the exact position of the aircraft on this ellipse could be found by taking a directional bearing on the echo signals. Under ideal circumstances the aircraft's position could be determined to within six miles, and the *Klein Heidelberg* erected by the German air force on the island of Romo, off the west coast of Denmark, could plot aircraft moving up to 280 miles away. However, by the time the device was fully operational there, Bomber Command had virtually ceased to route its aircraft over southern Denmark, and this unusual device had little effect on the night battles.

In the meantime, the Germans redoubled their efforts to jam the British *H2S* and 'Oboe' bombing aids, but they met with only moderate success. The *H2S* airborne radar carried

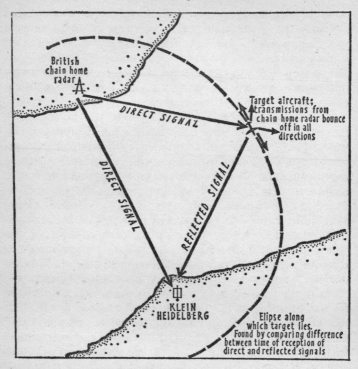

Klein Heidelberg

by the bombers presented the Germans with the most difficult problem of all, and they never abandoned the idea of
erecting metal decoys designed to give the same response on
the *H2S* screen as did a real city. The notion was not unlike
the launching of the ghost 'fleets' during the D-Day invasion;
but while it was one thing to design a radar-reflector capable
of returning an echo like a ship, it was quite another to make
a reflector produce an echo like a city. The Germans began
by setting out scores of metal tetrahedrons in the vicinity of
important towns: in this way expanses of open countryside
could be induced to give strong radar echoes like those from
built-up areas. Similarly, areas of water could be adorned

with radar reflectors mounted on rafts. The idea failed because too many reflectors were required to produce a convincing result, and each reflector had to be constructed to impossible tolerances. For example, to spoof the Mark III version of *H2S*, which operated on an even shorter wavelength than the original version, five hundred radar reflectors were needed for each square mile; each reflector had to measure nine feet across, and each face had to be perfectly flat and its angle to the other faces correct to one-third of one degree. Half-hearted attempts were made to use reflectors to alter the distinctive radar shape of certain towns like Kiel and Wilhelmshaven, but in each case too few were used and RAF crews reported no difficulties that could be attributed to the reflectors.

The jamming of 'Oboe', the RAF's precision blind-bombing aid for the Ruhr, presented similar difficulties. We have seen how during an 'Oboe' attack on Rheinhausen at the end of December 1943, the Germans had suddenly found the best means of jamming the original 'Oboe', the Mark I, causing the complete failure of the raid. Thereafter, the RAF had brought into service a Mark II and a Mark III 'Oboe', which differed from the original system in that they worked on centimetric wavelengths, which were very much more difficult to jam. On top of this the original Mark I 'Oboe' transmissions were kept up, to give the Germans something to jam. During March, Reichsminister Speer, the Minister of Munitions, had called for the fullest reports on this 'Oboe' system which was bringing such destruction to the Ruhr; Speer authorized his own high-frequency electronics expert, Professor Lüschen, to give maximum support to the drive to find countermeasures to 'Oboe', but the other experts still doubted that centimetric technique was being used. It was not until 3rd July 1944, during an 'Oboe' attack on the Scholven synthetic oil plant, that 'Oboe' type signals were clearly picked up on the nine-centimetre wavelength, four full minutes before the original Mark I signals were picked up. Now the Germans realized the truth: the latter had only been used these last five months as a camouflage for the former: on this one occasion, the RAF ground transmitter had switched the camouflage on four minutes too late.

The success of the invasion of France had resulted in a complete change in the conditions under which Bomber Command attacked its targets. At the beginning of August 1944, the first armoured spearheads had broken out of the German ring of steel surrounding the Normandy beachhead, and within a month the greater part of France was in Allied hands.

Previously, the area of the Franco-German frontier along the line of the Rhine had been of little significance during the RAF's night operations, and the Germans had expended little effort on building up any aircraft reporting system in this region. The loss of France left a gaping hole in the German chain of early-warning radar stations. From September 1944 onwards, almost all the RAF's bombing attacks were in consequence routed into Germany over France. While the German air force toiled to reposition radar sets to plug this breach, an even more serious tactical problem arose: by September, the air force's fuel supplies began sharply to diminish as a result of the Allied attacks on oil targets begun four months before. All training stopped, and after a time all non-operational flying was curtailed altogether.

It was against this grim background that RAF Bomber Command's specialist Radio Countermeasures unit, No. 100 Group, began to operate in strength over Germany. The new Group's functions were to fulfil two primary tasks: firstly it was to jam the German radar and communications channels, and secondly it was to provide large-scale night-fighter support for the bombers to screen them against enemy fighter attacks. The new Group was to 'confound and destroy'.

Ever since August 1943, Air Chief-Marshal Sir Arthur Harris had recognized that the respite his force was to gain from the introduction of 'Window' would be but a brief one. Bomber Command's predicament could be likened to that of a swimmer, treading water: they had to keep moving just to stay where they were. One major technical breakthrough in the Germans' equipment for night-fighting – and night-bombing operations could become impossible. Air Marshal Saundby, who had been Harris' deputy commander since earlier that year, later said: 'We were always worried lest the German scientists should make some technical breakthrough that would dramatically increase the effectiveness of the defences. We continually took up every device in order to prepare at every point – so that if the Germans did come up with any technical breakthrough we hoped that one or other of our gadgets could be modified to encounter it.' The new German radar, *SN-2*, was a case in point: to enable the Mosquito night-fighters operating over Germany to home on to the emissions of *SN-2*, RAF technicians produced a modified version of 'Serrate' to cover its frequency. The introduction of this device coincided with that of another which made use of the German aircraft's own emissions, code-named 'Perfectos'.

'Perfectos' was one of the neatest electronic gadgets to come out of the Second World War: it alone gave all three pieces of information necessary for a successful interception – direction, distance and positive identification. The principle was sure and simple: 'Perfectos' sent out a string of pulses in such a way as to trigger the identification sets of all German aircraft in the area. The German sets replied with special coded pulses, which not only betrayed the contact on the 'Perfectos' screen as a hostile, but also gave an accurate range and bearing on it. But for all its elegance, 'Perfectos' achieved little tangible success. The German fighter crews engaged the simplest countermeasure to it of all – they switched off their identification equipment. This, on the other hand, had the effect of complicating still further the job of the German air force fighter controllers. They now had no positive means of telling friend from foe on their radar screens.

Meanwhile, the range of jamming devices necessary to cover the frequencies used by the Germans was becoming larger and larger, and the need for specialized jamming aircraft became clear – aircraft which would carry jamming equipment and no bombs. During the early discussions on No. 100 Group, it had seemed that some thirty aircraft would be able to give the degree of support needed; but at the end of September 1943 Air Chief-Marshal Sir Douglas Evill had presided over a conference held at the Air Ministry in London to thrash out the whole problem and it was here that it had been realized that nothing less than a unit of Group strength – a hundred aircraft – would suffice; it had also been pointed out that long-range night-fighter squadrons would be of great value for the support of the bomber force, and four squadrons were allocated to the new Group. No. 100 Group had been formally established on 8th November 1943 under the operational control of Bomber Command. E. B. Addison, later promoted to Air Vice-Marshal, was appointed to command the Group; once again, as in 1940 when he had commanded No. 80 Wing during the battle against the German radio beams, he had found himself charged with forming a specialist unit from scratch.

It was not until mid-1944 that No. 100 Group was fully operational. Addison had set up his Group's headquarters at Bylaugh Hall in Norfolk, and gained control of a number of squadrons. There was No. 515 Squadron, Fighter Command's jamming unit, which was in the process of re-equipping from Defiants to Beaufighters, and there was the veteran No. 141 Squadron, which was changing from Beaufighters to Mosquitoes. Two more squadrons, No. 169 and No. 239, each had a few Mosquitoes but no adequately trained crews. No. 214 Squadron, intended as a heavy jamming unit with Flying Fortresses, had started off with no aircraft at all. The only operational unit had been No. 192 Squadron, a 'ferret' unit which incorporated No. 1473 Flight, one of whose Wellingtons had so dramatically resolved the mystery of *'Emil-Emil'* in 1942; No. 192 Squadron was equipped with Halifaxes, Wellingtons and Mosquitoes. With its motley collection of aircraft and units, No. 100 Group was not a promising battle formation as it stood; yet Addison moved quickly to get his machines fit for operations.

Early in 1944, the first Flying Fortresses destined for No.

100 Group had arrived at the Scottish Aviation Company at Prestwick, and here men worked on them day and night under great secrecy to modify them for their new role. Mufflers riveted to the exhaust pipes screened the bright exhaust flames, for those bombers were to fly by night with the RAF. Their noses sprouted bulbous blisters to house the *H2S* scanners, and the bomb bay was sealed up to house the jamming gear. Initially, each Flying Fortress was fitted with eight 'Mandrel' transmitters, each set to a different frequency in order to cover the whole of the spectrum used by the German *Freya, Mammut, Wassermann* and *Jagdschloss* radar sets; each Flying Fortress would also carry an 'Airborne Cigar' installation – three transmitters – to obliterate the German air force's VHF communications channels. The Flying Fortresses were each allocated two extra crewmen as special equipment operators to work the jammers.

At the same time, the RAF was just bringing into service its most mighty airborne jammer yet, code-named 'Jostle IV'. Dr Robert Cockburn later described how he came to secure the production of this jamming transmitter. The success of the early jamming operations in support of Bomber Command had won for the Telecommunications Research Establishment's jamming section better and better priorities: 'Soon,' said Cockburn, 'I was riding the crest of a wave. I was on various committees, and if I said something was good it usually got in. Eventually it became perfectly obvious that we needed a high-power communications jammer. It was easy to write down on a piece of paper how much power was needed. My contacts with industry were good and they were a little short of work.

'The next thing I knew was that Metropolitan-Vickers asked me to go and look at this jammer they had built. I was thinking of something about the size of a large biscuit tin, but by the time all the protective devices, the pressurization, the aerial insulators and all the rest were in, "Jostle IV" really was a damn great thing. I was scared out of my wits, because I suddenly realized that no paperwork had passed about this at all. They had done it entirely on my say-so. Then I got a most angry letter from Headquarters – I was given an Imperial rocket. Did I realize that there had been no authority, no finance?' It must indeed have been disconcerting for Cockburn's superiors to find that here suddenly was a half-million pound contract being fulfilled – about which they knew nothing.

Whether or not the 'Jostle IV' contract had passed through unorthodox channels, by the time the transmitter appeared in August 1944 it was a most formidable apparatus: it was able to blot out simultaneously the whole spread of frequencies from 38 to 42 megacycles – the spread previously covered one frequency at a time by the lower-powered jammer, 'Airborne Cigar'. The target was the channels used by the German air force for VHF radio-telephone communications with night-fighters. To cover this frequency-spread with adequate power, 'Jostle IV' radiated two thousand watts of jamming – the most powerful airborne transmitter of any type then built. The main unit was a cylinder the diameter of a large dustbin and twice as high, weighing 600 pounds. This bulk, and the requirement for additional powerful electric generators, meant that the jammer could be carried only by the specialist jamming aircraft. The Flying Fortresses of the RAF's No. 214 Squadron were fitted with the device in place of their previous 'Airborne Cigar' installation. Within a very short time, the new jammer forced the Germans to abandon the use of VHF communications almost completely, and control of the enemy night-fighters reverted to transmissions in the three to six megacycle frequency spread; but this was already the target for 'Tinsel', 'Corona', 'Dartboard' and 'Drumstick' jamming.

In May 1944, No. 199 Squadron had also joined No. 100 Group, equipped with Stirlings; each of these bombers had also been fitted with a battery of 'Mandrel' jammers. The two specialist jamming squadrons then lay low, doing little before the D-Day invasion; it was vital that the Germans should have no inkling of the weight of jamming being arrayed against them. Once the invasion was over, they took up their designated task of bomber support: the support

took the form of a 'Mandrel' screen similar to the screen
set up by the Defiants of No. 515 Squadron eighteen months
before.[3] The jamming centres in the screen were fourteen
miles apart, and formed a line running parallel to and some
eighty miles from the frontier. With the 'Mandrel' screen
running, all movements coming from behind it would remain
obscured from the German early-warning radar system. So
great had been the expansion of this German system's frequency cover that two four-engined bombers were now
required to orbit each jamming centre, to do the work one
small converted fighter had been doing eighteen months
before. During August, the Group put up 'Mandrel' screens
on sixteen different occasions.

To distract the German defences still further, No. 100
Group introduced 'Window' spoofs in support of the bombing operations. Not unlike the ghost 'fleets' produced for
the D-Day operations, each spoof would employ up to
twenty-four aircraft flying in two lines of twelve aircraft
abreast of each other, with $2\frac{1}{4}$ miles between each aircraft
and one line thirty miles behind the first. Each aircraft
released one bundle of 'Window' every two seconds. In this
way the formation would sow a radar echo-reflector the size
of a bomber stream of some 500 aircraft. It was for the
German fighter-control officers to divine whether the
echo came from just 'Window' or 'Window' plus five
hundred bombers; since the real bomber stream also continually released 'Window' the problem was a very real
one.

Air Vice-Marshal Addison had expected that these spoof
formations would suffer heavily, for they were only a small
force of aircraft and they had to draw upon them the wrath
of the defenders so that the real bombers could go unscathed.
In the event, their losses were no higher than those of other
Bomber Command units, although the feint operations often
drew large numbers of fighter aircraft towards them. This
was because the area surrounding the spoofing aircraft was
so saturated with 'Window' that the German night-fighters
had great difficulty in finding them, and moreover the air-

3. See page 124 above.

craft were in very widely dispersed formation besides. Gradually No. 100 Group became more daring with these spoofing operations, and they began to penetrate deeper and deeper into enemy territory.

Throughout the summer of 1944, the number of four-engined aircraft available to the Group increased. No. 223 Squadron joined the Group with Liberators, and No. 171 with Halifaxes; and No. 199 Squadron had its old Stirlings replaced with the higher-performance Halifaxes. While the ghost 'bomber streams' emerged from behind the 'Mandrel' screen and sent the night-fighters haring after them, the real bomber stream would be coming up somewhere else, escorted by the Flying Fortresses and Liberators blanketing the ether with the powerful 'Jostle' radiations. In the autumn, the effectiveness of these 'escort' machines was still further enhanced by the fitting of 'Piperack', a jamming transmitter designed to blind the German *SN-2* night-fighter radar. No. 100 Group mounted 'Mandrel' screens and 'Window' spoof operations almost every night regardless of whether there were to be real bombing operations: in this way, the force maintained a steady pressure on the already over-extended and strained German defence organization, and its dwindling high-octane fuel reserves.

The Germans circularized all their *Jagdschloss* radar stations with detailed instructions on how best to operate in the face of the RAF's 'Window' jamming and spoofs, in the middle of September 1944. The radar operators were warned that the RAF was attempting to force the Germans to scramble their fighters too early, in order to weaken the defences for when the real raids took place later the same evening:

> The correct control of our fighters is possible only when we are in possession of a clear picture of the air situation, which means the faultless tracking of all formations. There is no doubt that this is made more difficult by the release of 'Window'.

Radar operators were advised to study closely the nature

of the traces left on the main range tube of their *Jagdschloss* screens. 'Window' clouds, they were told, quickly fell behind the head of the enemy formations, and a long track containing numerous 'blips' not unlike a large bomber formation resulted; after a few minutes, a trace looking like a caterpillar would be left, while the spoof's trace would be more like a number of 'small surf waves'; during this phase, the head of the enemy formation could be clearly seen, and in the same way any change in the formation's course could be determined very quickly. After about ten minutes, the original 'Window' would have begun to settle and it would be possible for an attentive operator to recognize aircraft targets in the 'Window' cloud; however, it was admitted that 'faultless tracking of an air target still offers great difficulties'.

Only after half an hour had passed would the 'Window' cloud have settled far enough for the *Jagdschloss* operators to plot single targets and bomber formations without error. For half an hour after the spoofing force passed, any estimate of the number of enemy aircraft present could be arrived at only by guesswork. This was the contribution of the 'Window' spoofs.

The destroying arm of No. 100 Group had also been strengthened during the summer, by the arrival of three more Mosquito squadrons, Nos. 23, 85 and 157; what was more important was that the two latter squadrons were equipped with the very latest in night-fighter radar sets, the American-designed AI Mark X. Most of the other night-fighter units in No. 100 Group were eventually to receive new Mosquitoes fitted with this superb radar, but re-equipment was a slow business. The combination of the best new aircraft with the latest radar and target-homing devices enabled the Group to make a decisive contribution to the decline of the German night-fighter force during the closing months of the war.

To prevent surprise attacks from the rear, the Germans fitted their night-fighter aircraft with an additional rearwards-looking aerial for the *SN-2* radar; the set's operator was able to select either forwards or rearwards cover by the movement of a lever. More *Naxos* sets were installed, and these also gave warning of attack by the radar-equipped Mosquitoes; by the end of the war, almost all German night-fighter aircraft had been fitted with *Naxos*. The policy of restricting the use of *H2S* had reduced the device's value for finding the RAF bomber stream, but it did pick up the transmissions of the Mosquitoes' AI Mark X radars. Even so, the Mosquitoes continued to take a mounting toll of the German fighters, destroying one or two each night; the importance of this apparently small number was that from November 1944 all German night-fighter units were required to fly at night in the ground-attack role, and only the finest of crews were held back to combat the Bomber Command attacks. Each German fighter crew knocked out by the Mosquitoes of No. 100 Group was therefore irreplaceable. Sometimes the crews would escape unharmed after being shot down, but this was not often the case; and the cumulative effect of the Group's operations was an erosion of the very foundations of the German night-fighter force. No longer could the night-fighters cruise at leisure about the skies of Germany, looking for bombers to bring down.

There can be no doubt of what the best of the German night-fighter officers were capable of if they could fight unhindered in the bomber stream: Major Heinz-Wolfgang Schnaufer had amassed 121 confirmed night-victories by the end of the war, far surpassing the 83 of Prince zu Sayn-Wittgenstein; Schnaufer's greatest success was on one day in February 1945, when he had shot down two bombers early in the morning, and then worked his way into the centre of the bomber stream that night and shot down seven more. His prowess as a night-fighter pilot may be judged from the fact that three of his overall score were bombers brought down with his upward-firing cannon, while he himself was actually corkscrewing below a corkscrewing bomber.

Air Vice-Marshal Addison's Group could not claim the sole credit for the sudden decline in Bomber Command's losses during the autumn of 1944 and the first weeks of 1945; but the unit's jamming, spoofing and intruding activities were undoubtedly responsible for ensuring that the Germans never did recover from the initial blow. An impressive tribute to the Group's work is to be found in the record of a conference held at the German Air Ministry in Berlin on 5th January 1945. General Galland, the Air Officer in command of the fighter force, recalled the great achievements of the night-fighters, but—

today the night-fighter achieves nothing. The reason for this lies in the enemy's jamming operations, which completely blot out ground and airborne search equipment.
All other reasons are secondary.

One expedient the Germans introduced early in 1945 was a small force of jet aircraft to deal with the RAF Mosquitoes which repeatedly attacked Berlin at night. Built up by Lieutenant Kurt Welter, this force employed the outstanding Messerschmitt 262 jet fighter, which was one of the few German aircraft capable of going fast enough and high enough to engage the British wooden bomber. Welter resuscitated the old *Himmelbett* system of close ground control for his jet fighters; this was possible, because the Mosquitoes did not operate in large enough force, nor drop enough 'Window' to embarrass the system. At present the German jets carried no radar at all, and they had to rely on the searchlights to provide illumination during the final stages of the interception. These lights were operated by women, and the slogan was 'Once the girls have got you, you've had it!' Unfortunately the axiom was as true for the German pilots as for the British, for as often as not the jet pilots found themselves blinded by the beams of the searchlights which were supposed to be helping them.

The jet-propelled anti-Mosquito operations began in earnest from Burg during January 1945; when Allied bombers made Burg, seventy miles west of Berlin, unusable, Lieutenant Welter and his pilots operated off a nearby stretch of straight Autobahn. But the weakness of these early jet fighters was their low endurance, which varied between

fifty-five and eighty minutes. And they were rather too fast
to operate effectively against the standard four-engined
bombers at night. Jet engines were very temperamental, and
if the pilot throttled back too far the engines were likely to
'flame out'. Welter later said that he was greatly helped by
the fact that the Mosquitoes attacking Berlin always used
one of three routes, the Northern, the Central or the
Southern; the Germans got to know them so well that they
referred to them as Platforms One, Two and Three. The
raids themselves were called the 'London–Berlin Express'.
His jet fighters probably accounted for most of the thirteen
Mosquitoes lost in the Berlin area during the first three
months of 1945.

This was one bright spot in an otherwise gloomy picture.

One by one Germany's remaining cities were falling in
ruins under a hail of bombs directed by a whole battery of
radar-aiming devices. The defending night-fighters were becoming increasingly more impotent; indeed, Bomber Command was launching daylight raids on the most massive scale
as well, escorted by hundreds of long-range fighter aircraft.
Bonn fell to a daylight formation-bombing attack delivered
by No. 3 Group using the *G-H* bombing technique, which the
Germans were still not jamming. Bomber Command was in
a position to deliver massive saturation raids using *H2S* and
the other well-tried radar techniques, and to strike effectively
at small precision targets like oil plants and communications,
even at extreme range. Now mobile 'Oboe' and *G-H* transmitters were following the Allied armies in their advance
across Europe.

Its most effective deep-penetration attack was delivered
in the middle of February, when the centre of Dresden was
destroyed in two terrible night attacks in rapid succession,
mounted from behind a 'Mandrel' screen set up by 100
Group aircraft flying behind the battle lines. No. 100 Group
flew 'Jostle' reconnaissance and signals investigation sorties
without losing a single aircraft that night, and the Group's
'Window' feints directed against the Mainz–Mannheim area,
and then against the Cologne–Koblenz area, were so convincing that they entirely deceived the German fighter con-

trollers. A German night-fighter pilot stationed at an airfield on the outskirts of Dresden wrote in his diary for that night:

13th February 1945:
   My saddest day as a night-fighter. Midday at the aircraft – *SN-2* adjusted. Evening, first scramble, naturally for A-crews only. Take-off too late. Huge firework display over the city. Jockenhöfer shot down by our own anti-aircraft guns. Then second scramble, rather before 2 a.m. No communication [by R/T or W/T] with divisional head-quarters [at Döberitz, near Berlin]. Apparently Division was in the dark . . . Result: major attack on Dresden, in which the city was smashed to smithereens – and we were standing by and looking on. How can such a thing be possible? One's mind turns more and more to sabotage, or at least a certain irresponsible defeatism among the 'gentlemen' up there. Feeling that things are approaching an end with giant strides. What then? Wretched Germany!

In the whole of Germany, only twenty-seven night-fighters were scrambled to ward off the mighty assault that night, and only nine of the 1,164 returning aircraft reported having been attacked. Five of the huge force attacking Dresden were lost, and of these one was brought down over the target by bombs falling from an aircraft above, and another crashed after a collision with a third near Frankfurt. The combination of 'confusion' and 'destruction' was at its peak. The attacks on Dresden were renewed by the American air force on the following day:

14th February 1945:
   Midday, alarm. Airfield evacuated. Bombers over Dresden and further bombing. If the Tommies see our overflowing airfield it's all up with us. How impotent we have become!
   At night more raids as expected. This time B-crews also scrambled, and in good time. Bombers' target, Chemnitz. But our sortie was under ill omen from the outset: inter-com. broke down, no radio beacon found, *FuG.16* [VHF

radio] jammed, picking up own flak, 'Window' jamming, and enemy warning radar devices. W/T communication with Prague cut off, so flew to south-west. Fired cartridges as distress and identification signals – saved in the nick of time by small satellite airfield at Windisch-Laibach. Made very short but succcessful landing . . . Very warmly received. Lucky this time – another fifteen minutes and we would have had to jump for it.

Even by day the RAF bombers ranged far and wide over Germany now, bringing havoc to the remaining targets of importance. During a spectacular daylight attack on 12th March, *G-H* radar technique was used by some bombers to deliver a blind attack on the central railway station at Dortmund through a pall of smoke caused by a bombing raid mounted by more than a thousand heavy bombers shortly before. By night, the air over Germany was full of 'Window' and jamming radiations of every description, and these made the work of the RAF's long-range night-fighters almost as hard as that of the German defenders. Typical of these operations by No. 100 Group's fighter units was that on the night of 16th March. Squadron-Leader Dennis Hughes, a flight commander in No. 239 Squadron, afterwards described how he had met the bomber stream as planned as it went out towards Nuremberg, the target, and he had left the bomber stream to starboard until the target was reached: 'Innumerable contacts on aircraft and "Window" in target area made selection of individual echoes very difficult and at times impossible. Many contacts were followed up, and in every case a Lancaster was identified. One fleeting visual on an aircraft believed to be a Junkers 88 was obtained, but could not be held visually, and its echo was completely obscured in very heavy "Windows" and other echoes.'

One of Hughes' German counterparts in the air that night, Major Werner Hoffmann, had taken off in a Junkers 88 from Erfurt-Bindersleben at 8.12 p.m. and had soon made contact with the bombers making for Nuremberg. While in the stream, he shot down three of the raiding bombers, but he lost the bomber stream soon after leaving the target area. Hoffmann decided to make his way home via the nearest radio beacon, '*Otto*'. Unfortunately for him, the British fighter-pilot, Squadron-Leader Hughes, had seen a flashing beacon at the approximate position of 'Otto' shortly before 10 p.m., and soon after he had seen four white fighter flares. The RAF pilot turned towards them, and soon made radar contact with an enemy aircraft five miles away and crossing ahead of him. Hughes found that the enemy aircraft was taking 'strong evasive action', and Hoffmann was indeed performing the violent corkscrew he always adopted when British intruder aircraft were about. With his Ross night glasses, the squadron-leader observed his quarry to be a Ju. 88. He opened fire at a range of 600 feet, about three minutes after Hoffmann had left the radio beacon, and strikes were seen from the cannon shells all over the German aircraft; a second burst of fire set the root of the German's starboard wing and engine on fire. The enemy aircraft turned to port and began to lose height; then it went into a vertical dive, and exploded in mid-air. The attack had taken Hoffmann completely by surprise, but he and his three-man crew were lucky: they abandoned the Junkers 88 through the belly-hatch and all survived the night.

If the fog of the radio war caused British and German night-fighter pilots difficulties, it finally broke the morale of the German ground-controllers altogether. When Bomber Command delivered a twin-attack on Witten and Hanau on the night of 18th March, General Schmid personally intervened to order the bulk of the night-fighters to Kassel, although his most experienced fighter-controller was convinced that the move in the direction of Kassel was only an RAF 'Window' spoof, as it was. During an RAF Mosquito raid on Berlin four nights later, the tactics of No. 100 Group succeeded in diverting no less than six squadrons of night-fighters, by means of a 'Window' feint against the Ruhr. The feints involved on average only sixteen aircraft, yet the Germans invariably estimated that 150 or 200 aircraft were involved in them.

*

In the spring of 1945, Bomber Command's tactics of deception and radio-countermeasures had reached a fine perfection – just how fine can be assessed from one typical night operation on 20th March 1945, which can be examined in detail. That night, 235 Lancasters and Mosquitoes set out from England to bomb the synthetic oil refinery at Bohlen, just south of Leipzig. This force's zero hour was set at 3.40 a.m. on the 21st. Almost simultaneously, 166 Lancasters headed for the oilfield at Hemmingstedt in Schleswig-Holstein, far to the north of the first target. This force's attack was to commence at 4.30 a.m., and together with the other attack involved Bomber Command's main effort. In the meantime, the evening's diversions began with a large-scale nuisance raid on Berlin: thirty-five Mosquitoes of the Pathfinder force bombed the city, beginning at 9.14 p.m.

Just after 1 a.m. the main Bohlen force crossed the English Channel on a south-easterly heading, while a few miles to the south a feinting formation, comprising sixty-four Lancasters and Halifaxes from training units, crossed the Channel on an almost parallel course. It was here that the complications for the German radar operations began, for by five minutes past 2 a.m. an eighty-mile long 'Mandrel' screen, comprising seven pairs of Halifaxes from Nos. 171 and 199 Squadrons, was in position over northern France, throwing up a wall of radar jamming through which the German early-warning radar could not see. Shortly after crossing the coast of France, the Bohlen force split into two streams, hidden behind the 'Mandrel' screen: forty-one Lancasters broke away and headed off to the north-east, and these were to cause considerable difficulties for the Germans.

While the bomber formations were still approaching the German frontier, fourteen Mosquito fighter-bombers of Nos. 23 and 515 Squadrons were fanning out in ones and twos, and making for the airfields the German night-fighter force was expected to use that night. Once there, they orbited overhead for hours on end, dropping clusters of incendiaries and firing at anything that moved.

At 3 a.m. the training aircraft which had by now almost reached the German frontier near Strasbourg, turned about and went home, their work done. At the same time, the two formations making their separate ways to the Bohlen refinery burst through the 'Mandrel' jamming screen. Seven Liberators of No. 223 Squadron and four Halifaxes of No. 171 Squadron went five minutes – about eighteen miles – ahead of the larger Bohlen force, laying out a dense cloud of 'Window' which effectively hid the bombers following them. Once over the Rhine, the more southerly of the two streams of bombers turned north-east straight towards Kassel. So far, there was no way in which the German fighter controller could tell the real target for the night. In fact, at the time the bomber forces crossed the German frontier, the German fighter-controller of the Central Rhine defence area, Major Rüppel, seriously under-estimated the strength of the two approaching formations: he thought each force involved about thirty aircraft, and both might well be 'Window' feints. As the reports from the ground observation posts began to come in, however, it became clear that the southernmost of the two was much larger than he had estimated: no amount of jamming could conceal the roar of eight-hundred aircraft engines.

In his massive concrete bunker at Dortmund, Rüppel pondered the possible target for the night. One force was heading for Kassel from the south, and the other might turn in from the north at any moment. The situation thus seemed reasonably clear, but that was often the case with spoof attacks as well. There was not much time for thought, either: if he waited for the situation to clarify still further the bombers might complete their attack and be away before his fighters could arrive on the scene. Eighty-nine night-fighters had been scrambled at the first signs of enemy bombers approaching. Now they were orbiting beacons near their home airfields, awaiting his instructions. The German controller decided they had to be committed now, and ordered the whole of his force except one squadron to go to the radio beacons '*Silberfuchs*', '*Werner*' and '*Kormoran*' to the south and west of Kassel. The remaining squadron, the second squadron of *NJG2*, he sent to beacon '*Otto*' near Frankfurt to cover the possible threat to that city.

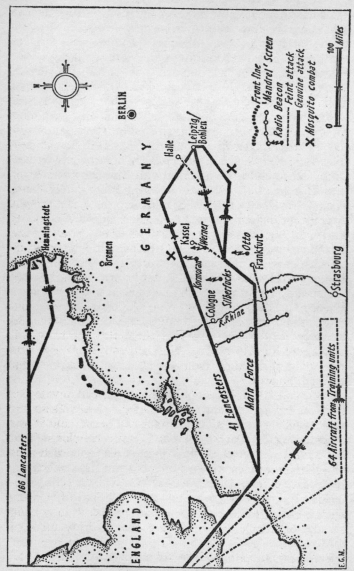

Bomber Command Operations 20th–21st March 1945

At first his guess on the bombers' intentions seemed to have been correct: from Kassel came reports of all the signs typical of the opening of a night-attack. At eight minutes past 3 a.m. brilliant flares began to blossom in the sky over Kassel, and the first few bomb detonations were heard. In fact, the 'Windowing' Liberators and Halifaxes had brought their feint right up to the city, and twelve Mosquito bombers of the Pathfinder Force had added to the effect with their flares and bombs, but these were not from the main force, and no further attack befell Kassel that night. Twenty-five miles from Kassel, the more southerly bomber stream had turned off towards the east, escorted by thirty-three high-level Mosquito night-fighters patrolling either side of them. One of No. 85 Squadron's Mosquitoes, piloted by Flight Lieutenant Chapman, picked up a Messerschmitt 110 on its 'Perfectos' homing device, and shot it down. At about the same time, one of the scores of German night-fighters orbiting near Kassel intercepted one of the 'spoofing' Liberators of No. 100 Group and shot it down; the heavy bomber plummeted down from 20,000 feet into a pine forest near Kassel. The sole survivor was one of the special operators for the radio jamming gear. He regained consciousness to receive a fright which, as he later put it, very nearly did kill him. The 'Jostle IV' canister, six hundred pounds of it, was lying on top of him. He suffered multiple injuries but survived the war.

The rest of this Liberator's crew had not died in vain, for the German night-fighters continued to wait over Kassel for some twenty minutes after the feint had begun, and it was not until 3.30 a.m. that the German fighter-controller realized that he had been tricked, and ordered his force eastwards after the bombers. Six minutes later the running commentary finally gave Leipzig, the nearest city to Bohlen, as the probable target, but by then it was too late even for the speeding night-fighters. As the German aircraft turned off in pursuit, the leading bombers were already within thirty miles of Bohlen.

Even now the spoofery was still not over. At 3.20 a.m. the larger of the two forces making for the refinery split yet again, and as the first target markers were falling on Bohlen at 3.40 a.m., other target markers were also falling on a target at Leuna of equal importance to the Bohlen plant, but some twenty miles to the north-west; four Flying Fortresses of No. 214 Squadron and two Halifaxes of No. 199 Squadron had laid a second 'Window' trail away from the main Bohlen force, and together with twelve Lancasters they marked out Leuna as if for a full-scale attack. Leuna was right in the path of the oncoming German fighters, and the spoof attack delayed them still further; one of the Lancaster crews involved paid the supreme price for the minutes of delay inflicted on the enemy.

It is in itself a remarkable evidence of the confidence with which bomber crews could now navigate to distant targets, that the Command could deliberately lay down all the marking and paraphernalia of an air attack on a target only a few miles from the proper one, and not have to fear that the main force bombers would be attracted by this marking themselves. The 211 Lancasters which now reached the main target, Bohlen, carried out a highly-concentrated eleven-minute attack upon the oil plant there, while five Flying Fortresses of No. 214 Squadron, a Liberator of No. 233 Squadron and two Mosquitoes of No. 192 Squadron gave the bombers jamming support over the very target. It was not until ten minutes past 4 a.m., as the last of the British bombers were leaving the target area, that the German night-fighters caught up with this force. The radar operators in each fighter aircraft encountered heavy jamming, and managed to bring down only two aircraft over Bohlen and three more during the bombers' withdrawal. The same Mosquito as had shot down a Messerschmitt near Kassel also brought down a Heinkel 219 just to the west of Leipzig, the second of the two German aircraft claimed by No. 100 Group's night-fighters this night.

But the night's operations were still not over, for seventeen minutes after 4 a.m. events began to boil up in northern Germany as well: twenty-seven Mosquitoes of the Pathfinder Force attacked the city of Bremen. While German attention was focused on this force, one-hundred and sixty-six Lancasters swept in low across the North Sea, below the horizon of the early-warning radar system – the second of the night's main forces, bound for the oil-plant at Hemmingstedt. The force kept at below 5,000 feet, maintaining strict radio silence until they were close to the target: shortly before getting there, they pulled up to 15,000 feet, bombed, and dived back below the cover of German radar again. A Liberator, a Flying Fortress and a Mosquito gave jamming cover during the actual attack, which opened at 4.23 a.m. and lasted for sixteen minutes. The Hemmingstedt raid passed almost unnoticed by the German raid tracking organization. It was certainly never appreciated as a major attack, and the bombers were already withdrawing when the first radar plots on 'weak formations' were reported. The sole loss was one Lancaster, which fell to a night-fighter near the target.

The bombers themselves claimed to have shot down two German fighters, bringing the total British claim for the night to four. German records show that they in fact lost seven night-fighters. What happened to the other three will probably never be known, but it is not difficult to speculate: a tired pilot, trying to bring his aircraft down quickly on a dimly-lit airfield patrolled by RAF intruders, might misjudge his approach and crash; a crew might return to base just a little too low for safety and fly into a hillside. A cautious German wireless operator might switch off his recognition equipment to prevent the RAF from homing on him with 'Perfectos', and he might be shot down by 'friendly' anti-aircraft fire. Such losses, which were frequent, were the results of No.100 Group's efforts just as surely as were those which the night-fighters actually shot down in combat.

Even so, there were signs in the closing months of the war that the Germans were beginning to overcome some of their difficulties. They were just mastering the menace of the small-scale nuisance raids by Mosquitoes in the last weeks of the war. The first Messerschmitt 262B night-fighters equipped with search radar were becoming available to Lieutenant Kurt Welter's force. During the five operational sorties that these aircraft were able to fly before the war ended, they claimed three Mosquitoes; but such successes would have become impossible the moment that the Mosquitoes brought radar-jamming aircraft with them.

A partial answer to the *H2S* airborne radar had also been evolved – a jamming transmitter which worked on centimetric wavelengths with sufficient power to affect the British radar; code-named the *Postklystron*, the device was effective only over short ranges however. And the only target to receive jamming cover by this means before the war ended was the vitally important oil refinery at Leuna. By this time, however, the *H2S* Mark III was in operation, and this worked on an even higher frequency than the earlier version; so once again it was for the Germans a case of too little, too late. By April 1945, when mobile transmitters were making it possible for 'Oboe' aircraft to bomb Berlin itself, the Germans had developed the means of jamming the Mark II and Mark III centimetric 'Oboe' with moderate success.[4] The German air force was on the point of introducing centimetric wavelength radar sets into large-scale service as the war ended: for the night-fighters there was *Berlin*, and for the anti-aircraft gunners there was a set called *Egerland* designed to replace the now hardly usable *Würzburg* equipment. For the harassed fighter-controllers on the ground there were the improved *Jagdschloss-Z* and *Forsthaus-Z* centimetric radar sets designed to replace the earlier *Jagdschloss* and other sets. In each case the prototypes had been built, but none existed in sufficient quantity to have any effect on the outcome of the war; within a few more months these would have been able to mount a formidable threat to the bomber forces once more, since neither the electronic nor the 'Window' jamming techniques then in use would have had much effect upon them.

Finally, the Germans were on the point of overcoming the greatest of all impediment to their night-fighter force, the jamming of ground-to-air communications channels. Like

4. The *G-H* transmitters were also brought up far into the heart of Europe, and were used on 14th April 1945, when Potsdam, near Berlin, was heavily bombed at the request of the Russians.

the *X-Gerät* bombing mechanism they had used in the earliest days of the war, their solution to this problem was quite sophisticated; they had evolved a system of ground-to-air communications code-named *Bernhard*. The system was difficult to tamper with. The *Bernhard* transmitters, which were erected throughout Germany in the closing weeks of the war, employed a large aerial structure, some seventy feet high and almost as broad, to focus the transmissions into a narrow beam which rotated once per minute. A growing number of night-fighter aircraft was being equipped with the corresponding receiver, *Bernhardine*; the receiver interpreted the system's coded signals and printed them out on paper tape in the manner of a teleprinter. A further valuable refinement was that the receiver also printed out the bearing of the aircraft from the ground station, and identified the station concerned with a code-letter. Every minute, as the narrow beam transmission swept round past the aircraft, the airborne *Bernhardine* receiver printed a new bearing or orders, or repeated the old ones. The information was presented in the form of a simple code, thus:

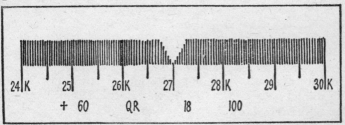

The *Bernhardine* picture

The bearing of the aircraft from the ground station was read, in tens of degrees, by taking the number seen at the apex of the 'V' (in this case 270 degrees, or due West). The letter printed every 20 degrees along the bearing scale identified the *Bernhard* station concerned: '*K*' indicated that the transmissions came from the ground station at Leck in Schleswig-Holstein.

The 'running commentary' on the movement of the bomber formation was transmitted in standardized form, and it was printed below the bearing information. The air situation reported in the example shown would be decoded as follows:

+ = *beginning of message.*

60 = *height of the leading aircraft in the bomber stream is 6,000 metres.*

QR = *leading aircraft in the stream at Fighter Grid reference QR (near Mainz).*

18 = *course of bomber stream (in tens of degrees), i.e. 180 degrees, or due South.*

100 = *estimated strength of bomber formation, 100 aircraft.*

Because it relied on beamed high-powered transmissions, and because teleprinter signals, like Morse code, had good 'break through' qualities, the *Bernhard* system would have been almost impossible to obliterate by jamming transmissions of the type put out by 'Jostle'. The system was scheduled to enter large-scale service in the late summer of 1945, but the Reich was overtaken by other events in May.

At midnight on 8th May 1945, the war in Europe came officially to an end. Allied forces had already overrun the greater part of Germany, and everywhere there was chaos and destruction. The High Command had decreed that the retreating troops were to leave nothing that might be of use to the invading enemy, and as the German armies had steadily fallen back the intricate defensive line upon which Generals Kammhuber and Schmid had lavished so much effort succumbed to the demolition charges. Even at the very end, however, there were still parts of Europe occupied by German forces which had yet to feel the impact of war. In Schleswig-Holstein and Denmark for example the air defences were still in working order. Moreover, it was here that many German air force units had ended their war, having retreated from all over Europe to these northern latitudes.

This presented the Royal Air Force and British Intelligence authorities with a unique opportunity to examine the now vanquished *Luftwaffe* organism. How effective had the British campaign of radio countermeasures been? Had No. 100 Group's night-fighters caused the enemy much trouble? What plans had the Germans made to regain the initiative in their night skies, which they had lost so completely in September 1944? What new equipment was coming into service? The war had been ended just thirteen days when two Halifax bombers touched down at the German fighter airfield at Jagel, near Hamburg. The aircraft brought an RAF investigation team, commanded by Air Commodore Chisholm – No. 100 Group's Senior Air Staff Officer – and consisting of senior officers from Bomber Command, Fighter Command and the Second Tactical Air Force.

It was not until the investigation officers arrived at an operational night-fighter station that they really learned the type of war the Germans had been forced to fight. At Eggebek airfield, the RAF officers met some of the German crews who had on occasion dealt such punishing blows to Bomber Command. Amongst them was Major Heinz-Wolfgang Schnaufer, of whom one of the team later reported: 'He wore a high-peaked cap suitably thumbed and twisted in true aircrew style and a mass of medals including the Knight's Cross with Oak Leaves, Swords and Diamonds – a magnificent affair that any dowager would be proud to wear on her corsage.' From the very start of the interrogations, it was clear that the Mosquitoes of No. 100 Group had caused the greatest embarrassment to the German crews. They had over-estimated the British fighters' capabilities, and were very much in awe of the combination of Mosquito aircraft with AI Mark X radar. The German pilot's prayer was said to be: 'Dearest Hermann, give me a Mosquito!' In their efforts to avoid the Mosquitoes' unwelcome attentions, the German fighter pilots had had to resort to some highly dangerous flying: pilots spoke of cross-country flights at less than 150 feet, dropping to 100 feet in the vicinity of their home airfields. One pilot, Captain Hans Krause, was wont to line himself up on his home runway at 10,000 feet, and then dive on it, landing at the end of his dive on an unlit airfield. This method, he said, had one main advantage: 'If you were shot down by a Mosquito, you had plenty of time to bail out!'

For the German night-fighter force, the coming of the Mosquitoes, and the great 'Mosquito panic' which had followed, had been the last straw. The 'Piperack' jamming had made the German night-fighters' airborne radar sets virtually useless: Schnaufer described how he used to make for the position where the jamming was strongest, and then search visually. After some prompting, he revealed a scheme he had devised, called Operation FEUERSEE. He had intended that almost the whole of his Wing, some sixty aircraft, should take part in this operation: they were to fly out low over the North Sea, towards a radio beacon near Orfordness. About fifteen miles short of the English coast they would turn southwards towards Ostend, a route calculated to cross that of the returning bomber stream. Low down over the sea, the maximum range of the fighters' airborne *SN-2* radar would have been at its best. Once the bombers had been picked up – it was known that they would by then have switched their navigation lights on to reduce the risk of collision – the German fighters were to climb and destroy them. The authorities were, he said, 'so slow in getting permission through the official channels that the war was over before it was granted'. Such a bold scheme would have worked only once, but it might have resulted in the destruction of a hundred bombers or so. The RAF interrogators admitted that this was so.[5]

On the subject of new equipment, the British investigators learned with some satisfaction that there was very little that they had not already heard about through their own Intelli-

5. Heinz-Wolfgang Schnaufer survived the war, but he was not to die in his bed. On 15th July 1950, he was driving his car on a business trip in France when a lorry pulled out from a side road across his path; in the resulting collision, the German driver was thrown out of his car, only to be killed when the load of compressed-air cylinders that the lorry had been carrying broke loose and rolled down on top of him. Major Schnaufer's accredited score of 121 night victories was never equalled.

gence channels. Apart from the *Bernhard* communications system just described, one device which did arouse considerable interest was the *Kiel-Gerät*, an infra-red homer which enabled night-fighter aircraft to close on to the radiations from a bomber's engine exhaust from up to four miles away. The equipment would see through a thin haze, but not through cloud. Both the moon and the conflagration at the target itself were liable to be an embarrassment to it, since these also emitted infra-red radiations; one of the German aircrew men spoke of having homed on to fires in the Dortmund area, from Kassel, a distance of over ninety miles.

One of those engaged in the examination of the German equipment was Wing-Commander Derek Jackson, the officer who had done so much to prepare 'Window' for service, and who had flown as radar operator in the flight-trials of the two German night-fighters captured during the war. There was a startling sequel to his trials with the *Flensburg* device, which had been designed to enable German night-fighters to home on to emissions from the bombers' 'Monica' tail-warning radar sets; it will be remembered that as a direct result of Jackson's trials in the summer of 1944, the dangerous 'Monica' sets had been withdrawn from service in Bomber Command. Now Jackson learned that the Germans had developed a ground direction-finder which also gave bearings on the 'Monica' transmissions. He was told that the device had been extremely useful to the Germans, as had the airborne *Flensburg*, until the beginning of September 1944 when for some reason that the Germans could not fathom the RAF bombers had ceased to emit these signals. The full significance of the use made by the Germans of British radar transmissions was brought home to the RAF investigators by one of the German aircrew who had fought on the Eastern Front. He said: 'Night-fighting was difficult there, because the Russians were so backward in radar that they had no transmissions on to which we could home!'

After this initial examination of the German air defence system, there followed a series of live exercises probably without parallel in the history of warfare. With RAF officers and men looking over their shoulders, German air force personnel once again manned a whole sector of the German defensive radar chain. Code-named POST MORTEM, the series of exercises were designed to discover exactly what the jamming had looked like to the Germans. The whole German air defence network in Denmark was to be employed for these, except for one section in the country's northern tip. Ten large radar stations with forty radar sets between them were included in the area of the exercises. All were linked by landline with the Fighter Division's control centre at Grove in central Denmark.

Operation POST MORTEM fell into eleven parts. Each of the radio-countermeasures tactics, and each type of jamming, were re-enacted by RAF aircraft to determine just how effective they had really been. The exercises took place in daylight to minimize the risk of collision, and for obvious reasons German fighter aircraft were not included in them. Major-General Boner, the Chief Signals Officer of the Reich Air Group – corresponding approximately to RAF Fighter Command – commanded the German contingent. The success of POST MORTEM would clearly require the fullest co-operation of the German radar operators and signals staffs who had until a few weeks before regarded the RAF as hated enemies. Perhaps surprisingly, the co-operation was forthcoming. As Major-General Boner explained, sooner or later the British were bound to clash with the Russians, who would slowly spread across Europe; Boner thought that a new war might break out in a few months' time. If Russia defeated Britain, Germany would in his opinion become a permanent extension of the Eastern bloc. If Britain won, it would be in her interest to have a powerful buffer-state between herself and Russia, and this circumstance would give Germany her greatest chance of revival. Consequently it was in the German interest to see that Britain was made as strong as possible in readiness for the coming war.

The first exercises opened on 25th June 1945, and the whole series was complete within little more than a week. They revealed many things, some of which were expected, but some of which were not. Most important was the proof that the British jamming campaign had obliged the Germans to use a system which could at best provide only a very loose form of control over the night-fighters. If the night-fighters' airborne radar could also be jammed, it was thus extremely difficult for any but the very finest crews to achieve useful results. The 'Mandrel' screen was not as successful as Bomber Command had hoped. Somewhere, a man at a radar set or in a listening post always seemed to come up with some accurate information on what was going on behind the 'screen' of jamming; it must be remembered that some of the German operators had had more than two years' experience of working in an environment of jamming. The only sure way to reduce the early-warning time was for the attacking force to approach at low level.

Even if the jamming did not entirely deprive the enemy of their early warning, it did make it extremely difficult for them to maintain a steady flow of accurate 'plots' on the bombers. During the final, and most elaborate, of the POST MORTEM exercises, the bombers had been accurately arrowed on the Grove situation map up until the time they crossed the western coast of Denmark; then, for the next forty minutes, the main force had been able to simulate an 'attack' on Frederika, in eastern Denmark, and re-cross the west coast without appearing on the German map at all. During this period, other plots had been confidently marked on the situation map: first there had been a 'Window' spoof, and then there had been a plot that can only have originated in the imagination of the fighter-control officer, for it bore no relation whatsoever to what was really happening. Furthermore, the German controllers almost always made a poor showing in their attempts to assess the size of formations when 'Window' was being dropped. On occasions the plots were under-estimated or exaggerated by factors as high as ten, and sometimes 'Window' was reported as being dropped when none was used at all. On no occasion – and this was important – had the Germans been able to tell with any degree of certainty which were real attacks and which were feints. One of the radar operators confided that for his job one needed to be something of a clairvoyant. And during

POST MORTEM exercise Number 8 the Grove situation map had shown a string of 'plots' on a force estimated at 150 bombers, when in fact there had been no bombers at all, only 'Window' dropped half an hour before by a small feint force.

With the end of POST MORTEM the RAF disarmament teams set about dismantling and demolishing the German radar stations. More than one hundred separate installations were removed to Britain, America and France for detailed examination. One of the radar sets which arrived at the establishment at Farnborough was a venerable old *Freya*, Serial No. 1, which had been built in 1936. A lot had happened since then.

Before we draw conclusions on the effect and value of radio countermeasures during the Second World War, it is necessary to consider the use of such techniques outside the strictly limited context of the RAF night bombing attacks.

Initially the daylight bombing attacks mounted by the US Army Air Force on targets in Europe had been confined to those days when the skies were clear. Since they lacked electronic aids, the bomb-aimers could hit their targets only under such conditions. Thus the German anti-aircraft gunners were able to engage the day bombers using their optical systems; they did not need to place great reliance on their fire control radars and so no great reduction in losses could be expected if these radars were jammed. Over Europe, however, it soon became clear that cloud cover at the target was the greatest single limitation to the effectiveness of daylight bombers. Even during the summer it was unusual if there were more than ten days in any month when the skies were completely clear over a target.

In the autumn of 1943, therefore, the US 8th Air Force in Europe also began using the 'Oboe' and *H2S* bombing systems pioneered by the Royal Air Force; and soon afterwards it introduced *H2X*, an American designed and built radar similar to *H2S*. Daylight raiding formations, too, began to operate with one or more pathfinder aircraft in the lead. If the bomb-aimer in the pathfinder aircraft could see the target during the bombing run, he aimed the bombs visually; if not, they were aimed using radar. Radar bombing would never be as accurate as visual bombing, but it was clearly a great improvement on any of the alternatives available to crews arriving in a target area blanketed by cloud: unaimed bombing; releasing the bombs on any reasonable target found clear of cloud on the way home; or turning round and taking the bombs home.

The daylight bombers could now attack cloud-covered targets with reasonable accuracy with radar, so it became necessary for the German anti-aircraft gunners to use their *Würzburg* radars if they were to engage the bombers. This, in turn, meant that now the daylight bombers could also benefit from the protection which jamming could give. Since the entry of the USA into the war, in December 1941, there had been a growing interest there in British work in the field of radio-countermeasures. Dr Frank Terman set up a research organization, the Radio Research Laboratory, at Harvard University. Working in close co-operation with his British counterparts, Terman's team drew up designs for several jamming devices. One of these, designated the APT-2 Carpet, was a noise jammer which covered the frequency band used by *Würzburg*; it was placed in full production.

From the autumn of 1943 the 8th Air Force heavy bombers began operating under the protection of 'Carpet' and 'Window'. During these operations it was found that the jamming transmitters complemented the protection from the metal foil. For a while the swirling clouds of 'Window' gave valuable protection for follow-up formations of bombers, there was no such protection for the aircraft in the leading formation. 'Carpet' extended this protection to the leaders, as well as providing additional cover to the aircraft coming behind.

From May 1944 the 8th Air Force was losing more aircraft to flak than to enemy fighters. But this was due more to the effectiveness of its own escort fighters in fending off the defending interceptors, than in any qualitative improvement in the German anti-aircraft fire. Flak was now the only effective method of defending targets against daylight attacks and, until the centimetric wavelength *Egerland* was ready, the *Würzburg* was the only effective method of blind fire control that existed. The band of frequencies used by *Würzburg* was progressively widened, until by the end of the war it ran from 460 to 580 megacycles. Just like Cockburn and his team in Britain, Terman romped through the frequency spectrum endeavouring to match each change introduced by the enemy. In its final form 'Carpet' comprised three separate jammers each able to cover any frequency in the band 475 to 585 megacycles, which the operator tuned on to the frequencies of *Würzburg* sets active in the area using a special receiver; in its general layout, the complete system was not unlike the British 'Airborne Cigar'.

One further daylight bombing operation deserves mention to complete this account of the European war. It took place not over Germany but over the north of Norway and Sweden; and although no jamming was involved it contained some interesting lessons for those wishing to consider the wider aspects of radio warfare. The target was the German battleship *Tirpitz*, anchored in Tromso Fjord; the attack was mounted by a force of thirty-two Lancasters of Nos. 9 and 617 Squadrons of the Royal Air Force on 12th November 1944.

Tirpitz was an extremely difficult target. She had to be attacked visually because radar bombing would not have been accurate enough; and the bombs had to be released from altitudes around 12,000 feet if they were to penetrate the thick armoured decks which shielded her vitals. For such an attack, surprise was essential. The Royal Air Force planners had to allow for the Germans having smoke generators positioned round *Tirpitz*; if there was sufficient warning, they could blanket the battleship with smoke and make visual bombing impossible. Moreover, situated 1,100 miles from the nearest airfield in Britain, *Tirpitz* was beyond the reach of the longest-ranging escort fighters. Forewarned, the German fighter units based around Tromso could have played havoc with such an attack.

As part of their general effort to follow the evolution of the German radar defences, radar reconnaissance aircraft of No. 100 Group had plotted the locations and arcs of cover of those stations situated along the Norwegian coast. The radars provided overlapping cover on aircraft approaching the coast at altitudes of 5,000 feet and above. Any attempt to jam these radars would alert the defences. The radar chain was less effective against low-flying aircraft, however; and over the central part of Norway there was a gap in the cover, through which aircraft flying at 1,500 feet and below might sneak unobserved.

Examining these features, the RAF planners decided on a route which offered a good chance of achieving surprise. The bombers were to stay below 1,500 feet and make for the gap in the radar cover over central Norway. After crossing the coast they were to continue almost due eastwards over the mountain range and into Sweden; then, keeping the mountain barrier as a screen between themselves and the prying German radar beams, the bombers were to approach *Tirpitz* from the landward side where they were least expected.

The attack went exactly according to plan. The bombers followed their briefed route and delayed their climb until the last possible moment. Reaching *Tirpitz* without any interference from enemy fighters, the Lancasters delivered an accurate attack and scored three hits with 12,000 pound 'Tallboy' bombs. The uncontrollable fires, started deep inside the ship reached the after magazine and the mighty explosion which followed tore a long gash down the side. *Tirpitz* capsized.

Afterwards Sir Ralph Cochrane, who as an Air Vice-Marshal had commanded No. 5 Bomber Group at the time of the *Tirpitz* attack, stated that it was the riskiest air operation with which he had ever been associated. If the battleship's defences had functioned as they could have, the attack would have been rendered a disastrous failure. In the event, however, the risks proved fully justified; *Tirpitz* was destroyed for the loss of only one bomber to anti-aircraft fire.

The use of evasive routing to achieve surprise was by no

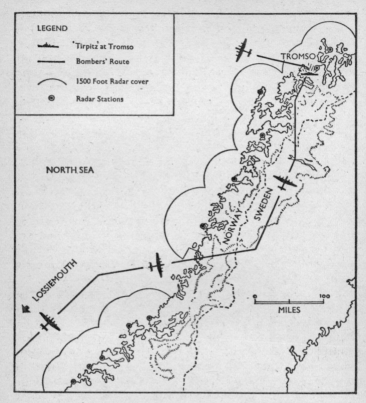

LEGEND

⚓ 'Tirpitz' at Tromso

—— Bombers' Route

⌒ 1500 Foot Radar cover

◉ Radar Stations

NORTH SEA

TROMSO

LOSSIEMOUTH

NORWAY

SWEDEN

100

MILES

Attack on the *Tirpitz*, 12th November 1944

means unique to the *Tirpitz* operation; by the end of the Second World War almost all air forces were using such tactics. The attack did, however, exploit to the full the possibilities of this course of action. First, there was the clever use of radar reconnaissance aircraft to plot the positions and cover of the German radar stations. Secondly, there was the skilful use of a low altitude approach to reduce the effective range of the radar stations. Thirdly, there was the use of a route which exploited the gap found in the enemy radar cover. Fourthly, there was the use of terrain, the mountain range, to screen the bombers from the enemy radar as they approached their target from the landward side. Fifthly, there was the use of neutral territory to enable the bombers to approach their objective undetected. From each of these standpoints the tactical planning of the attack on *Tirpitz* was exemplary and a clear pointer to the methods available to strike aircraft penetrating defences three decades later.

In the Pacific theatre, as in Europe, the use of radio-countermeasures had become firmly established by the end of the Second World War. Here, however, the Japanese radar systems proved quite unable to cope with the barrage of jamming thrown against them. In 1945 Japanese radar technology was at roughly the same level as that of the Germans had been in 1941; airborne radar for night fighters was still in the trials stage and the early warning and fire control radars were relatively crude. Against this rickety defensive system the USAAF hurled its massive jamming capability built to 1945 standards of technology.

Early in 1945 the B-29 Superfortress bombers of the XXIst Bomber Command began a series of intensive night attacks against Japanese cities. Each bomber carried one or two jammers to counter the defensive early warning and fire control radars, plus some 500 pounds of 'Window'. 'Porcupine' B-29's (so-called because they literally bristled with aerials) provided general cover for the raiders; these aircraft carried up to fourteen separate jamming transmitters and a ton of 'Window'. In the face of such a concentrated countermeasures barrage, the ill-equipped Japanese radar defences simply caved in. During the seventeen maximum-effort attacks mounted against Japanese cities, between 9th March and 15th June 1945, the B-29 heavy bomber units flew nearly 7,000 sorties; only 136 bombers were lost, or 1·9 per cent of the sorties flown. Moreover, many of the aircraft included in this figure had fallen not to enemy action, but as a result of mechanical failures or straightforward flying acci-dents. This low loss rate is the more remarkable when one considers that, in terms of anti-aircraft guns, the

Japanese cities were by no means weakly protected; the great conurbation of Tokyo, Yokohama and Kawasaki, for example, was defended by more than 500 anti-aircraft guns.

At a quarter past eight on the morning of 6th August 1945 a lone B-29 dropped a single atomic bomb on Hiroshima. Three days later the city of Nagasaki was similarly attacked. For the Japanese people, near to the end of their tether even before they had been hit by this most destructive of weapons, this was the end. On 15th August they surrendered.

What effect did the various radio-countermeasures have on the course of the Second World War? The initial achievements are easiest to assess in simple terms. Without doubt the success of No. 80 Wings' jamming attack on the German navigational beams played a vital part in Britain's survival during the hard winter of 1940. Had No. 80 Wing failed, the German air force might have succeeded in destroying both the will and the means of the British nation to fight on alone. This was unquestionably the campaign's greatest achievement of the war.

Next in importance was the success of the invasion of France, which owed much to the jamming effort. Without this support the fight to secure the beachhead in Normandy would certainly have been far bloodier.

The value of the jamming support given to the Royal Air Force bombing offensive on Germany is difficult to quantify. Because it is impossible to prove a negative in such a case, one cannot say for certain which bombers would have been shot down had there been no jamming and which would have survived. It seems likely, however, that overall the various jamming methods reduced Bomber Command's loss rate by just over one per cent, from December 1942 until the end of the war; this meant a saving of more than 1,000 bombers and their crews. It is relevant to mention that it took RAF Bomber Command from the beginning of the war until the summer of 1944 to bring its operational strength up to 1,000 aircraft.

Without the distraction of the jamming, the German defenders could have dealt far more severely with the night raiders.[6] It is, therefore, no exaggeration to say that the radio-countermeasures campaign played a major part in keeping Bomber Command's losses down to a level which the RAF could afford to countenance.

In the case of the American daylight attack on targets in Europe, it was the German fighters making visual attacks that inflicted the most serious losses. Radar played only a minor part in these, so radar jamming alone could provide little relief. Against flak it was a different matter; where cloud or (over Japan) darkness forced the German and Japanese gunners to rely on radar controlled fire the jamming could and did greatly reduce its accuracy. It is probably true to say that jamming cover saved between 300 and 500 American bombers and their crews.

How was it that the Germans came to find themselves continually 'trotting after' in the radar race? In truth, they were the victims of circumstance. From 1936 till 1942 Britain, always on the defensive, had little alternative but to concentrate her finest scientific brains on the development of radar. At first she desperately needed an effective radar warning chain to enable her outnumbered fighter squadrons to be used to maximum effect. The Battle of Britain almost over, the priority shifted to quite different radar equipment to enable night fighters to engage the German night raiders. Simultaneously the radio jamming organization was hastily formed, again at the highest priority. No sooner had the night Blitz ended than completely new radar equipment was again needed, this time to enable ships and aircraft to hunt the U-boats which were gnawing at Britain's vital sea communications. The year 1941 ended with demands from RAF Bomber Command for

6. This happened over Britain at the beginning of 1944 when the German bombers, lacking effective jamming support, lost on average one aircraft and four trained crewmen for every five British civilians killed in the attacks.

a completely new family of radar devices, to aid its crews to strike accurately. Each of these steps demanded new advances in the 'state of the art' from the scientists engaged in radar research; and this field enjoyed the highest priority when it came to the allocation of scarce manpower, materials and production capacity.

The Telecommunications Research Establishment mushroomed during this period, to enable it to meet the needs of the fighting services. From 1941 this massive research effort was coupled to the United States' formidable capability to develop and produce such equipment. When the Allies moved on to the offensive, radar had advanced to the point where it was able to assist these operations also.

Compare this forced-march with the dawdling pace of radio and radar development in Germany, which in four years allowed that nation to fall from a position of parity to one of patent backwardness. Until the winter of 1941 the German armed forces were continually on the offensive; radar was of little use to them and they made few demands on their scientists. The only advanced radio devices used on a large scale were the navigational beams. Two out of the three systems had been fully developed prior to the war and the third was not pushed as hard as it might have been. From 1941 the German armed forces had to focus their main attention on the Eastern Front; radar could play little part in the bloody war of attrition then being fought out by the opposing armies. When, in the summer of 1943, the German High Command came to realize their plight due to the backwardness in radar, it was too late. The armed forces and industry were combed for radio technicians,[7] but Dr Hans Plendl could do little more than prevent the technological gap

7. It was even necessary to recruit skilled labour in the concentration camps. In December 1943 Plendl reported to Heinrich Himmler that a high-frequency electronics research laboratory had been built by the SS in Dachau concentration camp, where one of the prisoners was Dr Hans Meier, the former director of the Siemens & Halske's Central Laboratory. Over a hundred skilled prisoners were to be employed on dismantling captured enemy equipment – the same sort of work Supper and his team were engaged in at Farnborough.

from widening still further. This unhappy state of affairs was aggravated by the work of Dr Cockburn and his team in Britain and Dr Terman and his team in the US, for the German research organization was forced to concentrate less upon new developments and more upon modifications of existing devices, as expedients to enable them to work in the face of the continually changing barrage of jamming. With the success of this jamming the morale of the German air force sagged and confidence in its electronic equipment withered.

When the war ended the German forces were on the point of introducing several powerful and novel weapons. It is sometimes argued that had the war gone on for another year the outcome might have been different. This notion does not stand the test of analysis; things move so quickly in modern warfare that such a gift of time canot be credited to one side while leaving the technology of the other barren. Germany did extremely well to hold out as long as she did; but in the face of the overpowering combination of Anglo-American technology, American production capacity and Russian manpower she was crushed, inexorably, like a nut in a vice. Only one weapon could have enabled the Germans to redress this unfavourable balance, the atomic bomb. But when the war in Europe ended the Germans were a long way from producing such a device, while the Americans were on the point of exploding theirs. Had the war continued for the conjectural extra year it is probable that the atomic bomb would have ended the conflict in Europe, just as it did so in the Far East.

The state of Japanese radar, when the American strategic bomber offensive on the home islands began to gain momentum in the spring of 1945, was so far behind that of the Germans that it posed few problems to those planning countermeasures; there were no significant improvements in Japanese radar in the offing when the war ended.

11

The Years of Shadow-boxing

> It may well be that we may, by a process of sublime irony, have reached the stage in this story where safety will be the sturdy child of terror, and survival the twin brother of annihilation.
>
> – SIR WINSTON CHURCHILL, speaking of Britain's nuclear deterrent force in the House of Commons in 1955.

Immediately after the end of the Second World War the science of radio countermeasures, upon which so much effort had been lavished, fell into disuse. At first there was no potential enemy against whose radars the countermeasures might be directed. Then, when the Russian threat did become clear, that nation's radar was at such a rudimentary state of development that countering it did not justify any diversion of the limited peacetime resources.

The first large-scale war fought after 1945 was that in Korea, which began in 1950 and ended in 1953. During the conflict the United Nations, but predominantly American, Air Force mounted bomber attacks on the Second World War pattern. Initially the B-29 Superfortresses were able to operate against both tactical and strategic targets with impunity by day. Following the entry of China into the war in November 1950, however, that nation's air force began to make its presence felt. Modern MiG-15 jet fighters appeared in action and these, backed by early warning radars located safely over the border in China, made life increasingly hazardous for the bomber crews. Finally, in October 1951, the threat of increasingly heavy losses forced Brigadier

General Joe Kelly, commander of the US Far East Air Force Bomber Command, to order his Superfortresses to attack only under cover of darkness.

It was the story of 1940 and 1941 all over again. At first the North Korean night defences were ineffectual and the bombers could range anywhere over the country with impunity. Then, gradually, gun and searchlight control radars began to appear at the more important targets. The radars were a hotch-potch collection; some of them were of US design, having either been captured from the Nationalist Chinese or else been copied by the Russians from sets supplied during the Second World War. There were no radar-equipped night fighters, so the defenders tried using day fighters to attack bombers illuminated by the searchlights; again, the parallel with the illuminated night fighting tactics of 1941 is clear.

During April and May 1952 B-29 crews saw enemy fighters operating at night on several occasions, but attacks were not pressed home. Then, on 10th June, the defenders did strike back. As four B-29's of the 19th Bombardment Group ran in to bomb the rail bridge at Kwaksan, more than twenty searchlights were switched on and held the bombers in their beams; then an estimated twelve jet fighters made repeated attacks. One B-29 exploded over the target, a second disappeared somewhere over North Korea and a third was damaged so severely that it only just made it back to an emergency landing ground near the front line. By itself, darkness was no longer an effective shield for the B-29's

The reverse stung the USAF into taking a renewed interest in jamming. Since the Second World War the name of the science had been changed from radio countermeasures to electronic countermeasures; but that was about all that had changed. Old Second World War jammers were taken out of storage, dusted down, retuned to the frequencies of the North Korean radars and installed in the B-29's. Despite the age of the equipment and the hasty training of the operators, this step paid off.

Indicative of the new-found respect for the North Korean defences were the tactics used during the attack on the Nansan-ni chemical plant, situated on the Yalu River border with China, on 30th September 1952. The raid was planned to coincide with a night when there would be good cloud cover over the target. The first to attack were three B-29's, which dropped shrapnel bombs to suppress the anti-aircraft defences before moving out to orbit positions nearby from which they jammed the frequencies used by the enemy gun and searchlight control radars. Forty-five B-29's then bombed the target while seven B-26 Invader attack bombers ran in at low altitude and engaged the ground defences. The cloud and the suppression fire, coupled with the use of Chaff (as 'Window' was termed in the USAF) and electronic jamming, effectively prevented the searchlights illuminating the bombers and there were no fighter interceptions. A few of the bombers were holed by anti-aircraft fire but none was shot down.

When the Korean War ended in July 1953, the B-29 bomber units had flown some 21,000 operational sorties by day and by night. Thirty-four of these aircraft had been lost in action: sixteen to fighters, four to anti-aircraft fire and twenty to other causes. The loss represented 0·2 per cent of the operational sorties. Losses had been held down by exploiting every weakness of the defences and by changing the tactics to meet each change made by the defenders. Heavily defended targets were attacked only on dark nights and often attacks were timed to take advantage of cloud cover at the target; in addition, the defences at the target were often engaged by attack aircraft.

Undoubtedly the jamming cover played its part in holding down B-29 losses in Korea. The official USAF history[1] afterwards stated that without electronic countermeasures support the B-29 losses might well have been three times as great. In terms of electronic countermeasures, however, the Korean conflict was merely an extension of the Second World War. The raiding forces operating over North Korea made use of jamming support, but this was provided almost

1. *The United States Air Force in Korea* by R. F. Futrell.

entirely with equipment which had been available seven years earlier.

By the mid-1950s the USAF, the RAF and the Russian Air Force all possessed long range bomber forces armed with nuclear weapons. The second generation of nuclear-armed bombers were jet-propelled and initially flew fast enough and high enough to stand a good chance of penetrating the enemy defences without the protection of electronic counter-measures.

The renewed interest in electronic countermeasures came only at the end of the 1950s, with the introduction into service of surface-to-air and air-to-air guided missiles. Compared with those which were to follow, the early guided missiles were naïve in concept and all were vulnerable to enemy countermeasures. To neutralize the missile threat to their deterrent forces, the major air forces all began to equip their nuclear-armed bombers with electronic counter-measures equipment. After a hiatus of some ten years, electronic countermeasures were back in fashion again.

By this time 'centimetric' wavelength radars were in large scale service everywhere, so appropriate jammers were necessary to counter them. Many of the new jamming trans-mitters employed the carcinatron, a French development of the late 1940s which has been as significant to centimetric wavelength jamming as the magnetron has been to centi-metric wavelength radar. The carcinatron generates 'noise' at very high power; by adjusting the voltages to the elec-trodes, the carcinatron can be made to generate noise on a target frequency or spread of frequencies.

Typically, a long-range bomber of the early 1960s carried devices to jam enemy early-warning, fighter control and missile control radars, the fighter control radio channels and the interception radars fitted to the fighters; to ward off radar-homing missiles Chaff could be dropped and to decoy away infra-red homing missiles there were special flares. To provide warning if the bomber was being tracked by enemy radar there was a passive warning receiver (a descendant of the Second World War 'Boozer'); and to enable the crew to observe enemy fighters attempting to move into a firing position, and avoid accordingly, there was a tail-warning radar.

One interesting device introduced into the USAF at this time was the 'Quail' radar decoy missile. Launched from the bomb bay of a B-52 Stratofortress bomber, 'Quail' had the appearance of a small aircraft with a wing span of just over 5 feet. It carried radar reflectors in the nose to give it the same appearance as the B-52 on enemy radar; to complete the illusion, the missile could carry jamming transmitters and radiate jamming in the same way as a real bomber. After launch the jet-propelled 'Quail' flew a pre-programmed course into the enemy defences, to engage their attention while the real bombers sneaked through unobserved.

The long-range bombers' tactics at this time were to fly as fast and as high as they could in groups, spaced widely enough to deny the enemy an easy target but closely enough to support each other with their jamming. Many of the bombers now carried long range air-to-surface missiles, which reduced to a minimum or even eliminated the time spent in the heavily defended area close to the target.

Throughout the 'cold war' both sides have made strenuous efforts to keep track of any change in the defences of the other. As we have already seen, effective countermeasures are difficult to devise unless the frequency and other details of the enemy radar are known. Hence the continuing importance of electronic intelligence. It was to collect such information, and other forms of reconnaissance, that the imaginative U-2 project was initiated.

From its conception in 1953, the Lockheed U-2 had been built without compromise for flight at very high altitudes. Designer Kelly Johnson had gone to great pains to keep the structural weight of the aircraft as low as possible; the metal skinning of the fuselage was so thin that a man could almost

punch a hole through it with his fist. Fitted with a thin high
aspect-ratio wing spanning 80 feet, the U-2A weighed just
under 16,000 pounds with a full load and could cruise at
altitudes around 65,000 feet on the 11,000 pounds thrust of
a J-57 engine. The later U-2B, carrying extra fuel in tanks
mounted on the wings, weighed 17,200 pounds and could
reach altitudes of above 70,000 feet on the 15,000 pounds
thrust of a specially modified J-75 engine. In itself such an
altitude performance was remarkable enough for an aircraft
of the mid-1950s (in August 1965 the official world altitude
record stood at 65,876 feet); but when this was coupled with
the operating range of over 3,500 miles of the U-2B, it becomes clear that the aircraft was a brilliant technical
achievement.

Between 1956 and 1960 the U-2's roamed far and wide
on reconnaissance missions over Russia, operating from the
airfields at Incirlik in Turkey, Peshawar in Pakistan, Bodo,
on Norway and Atsugi in Japan. The Russians were, of
course, able to track the high-flying U-2's for long periods
on radar. But there is a world of difference between seeing
an aircraft on radar and shooting it down; the fact that these
aircraft were able to spend as long as seven hours over
Russian territory, watched by the alerted defences which
were unable to engage with either fighters or missiles, em-
phasizes the genius of Johnson and his design team. During
the late 1950's the Russians made several general protests
about 'Imperialist spy flights infringing Russian airspace'.
But they could not say how blatant these were without revealing to the world the fact that they were helpless to deal
with them.

As well as photographing military and industrial installations the U-2 carried special radar receiving equipment
feeding tape recorders, to gather information on any radars
radiating as it cruised high overhead. What can be learnt
from intercepted radar signals? Firstly, there is the frequency
of the radar, the most important thing to be known if it is to
be countered; if the frequency is known intelligence officers
can deduce how accurately targets can be plotted, how well
the set can see through cloud and also gain a fair idea of the

set's performance against low-flying aircraft. Secondly, from the rate at which the radar beam is made to scan through an aircraft, it is possible to deduce the purpose of the radar: whether it is an all-round-looking search radar, a height finder, a rapid-scanning radar of the type used for missile control, a locking-on radar associated with beam-riding or semi-active-homing missiles, and so on. Thirdly, from the rate at which the radar pulses are transmitted, it is possible to estimate the maximum usable range of the radar. Fourthly, from the time 'width' of the radar pulses, it can be learned how well the set can discriminate between several aircraft flying close together (or one aircraft flying through several Chaff clouds). Fifthly, from the location of the aircraft itself as the signals are picked up, it is possible to get a rough idea of the position of the radar; it is possible to determine those areas of high radar density where the defences are strong, as well as those of low radar density where the defences are weak. The list given is not comprehensive but it does enable the reader to gauge the importance of this form of reconnaissance. The old German saying 'All radio traffic is high treason' is as true as ever.

The U-2 flights over Russia went on unhindered until 1st May 1960. Early that morning one of these aircraft took off from Peshawar. Its pilot, Francis Powers, had been briefed to fly a zigzag track which would take him across Afghanistan, then over the important Russian cities of Sverdlovsk, Kirov, Archangel and Murmansk, then across northern Finland and Sweden to Bodo in Norway. The 3,300 mile flight was to be made at altitudes around 70,000 feet. At a ground speed of about 450 m.p.h. this meant that Powers would have to spend some eight hours in his cramped cockpit, dressed in his uncomfortable but necessary partial-pressure suit; not the least remarkable factor in the U-2 reconnaissance flights was the physical demands they placed on the pilots involved.

Almost exactly four hours after take-off Powers was nearing the city of Sverdlovsk, some 1,300 miles inside Russia. Apart from a malfunction of his automatic pilot the flight was going according to plan and, far below, the ground unrolled with scarcely a whisp of cloud to hinder the photography. Then, suddenly, there was what Powers later described as a 'dull thump'. The U-2 jerked forwards and a bright orange flash lit up the sky. A ground-launched guided missile had exploded very close to the tail of his aircraft. The nose of the U-2 fell, then it went into an inverted spin. Powers succeeded in struggling out of the falling aircraft and parachuted safely to the ground; soon after landing he was picked up by civilians who handed him over to the police.

The following August Powers went on trial in Moscow. The Russian prosecutor stated that the American pilot had '... photographed the territory of the Soviet Union for the purpose of espionage'. Moreover, examination of the tape spools taken from the wreckage of the U-2 had revealed that '. . . the signals recorded belonged to ground radar stations of the anti-aircraft defences of the Soviet Union'. Powers was sentenced to be 'confined' for ten years, the first three of which he was to spend in prison. In the event, however, he was exchanged for a Russian spy a year and a half later.

In considering the ability of the long-range bombers of the time to reach their targets at high altitude, it is important not to exaggerate the importance of the destruction of Powers' aircraft. The lone U-2 had received the undivided attention of the entire Russian air defence system that May morning; there is evidence that several fighters had attempted to intercept it and several missiles had been fired at it, before the successful shot exploded nearby. The remarkable fact about the reconnaissance flights is not that Powers' aircraft was brought down, but that the U-2's were able to operate for so long without any protection from electronic countermeasures before one was shot down. While the long-range bombers could not reach the same altitude as the U-2, they would penetrate the defences in large numbers and they had the powerful advantage of electronic countermeasures for their protection. The fact that a single non-jamming aircraft could be shot down at high altitude certainly did not prove that a large force of bombers flying at a slightly lower altitude with the protection of jamming could be prevented from destroying its targets.

During the years immediately following the destruction of the U-2, however, the deployment of large quantities of new defensive equipment increased the vulnerability of the high-flying bomber. Supersonic fighters fitted with radar were now entering service in large numbers. Moreover, to reach their targets, the long-range bombers now had to penetrate successive belts of surface-to-air missile batteries; many of the latter had been re-equipped with second generation weapons, which had been designed to operate in the face of jamming.

Yet, while these moves made things hotter and hotter for bombers endeavouring to reach their targets at high altitude, it became clear that if the bombers went in at altitudes of 5,000 feet or below the new weapons became progressively less effective. By picking their way round the radar cover, as the *Tirpitz* raiders had in 1944, the bombers were able to present only fleeting targets to the radar-directed fighters and missiles. The swing from high altitude to low altitude penetration tactics took place around 1963. The long range bombers of all major powers, the American B-47's, B-52's and B-58's, the British Valiants, Vulcans and Victors, the Russian Tu-16's (NATO code-name *Badger*), Tu-20's (*Bear*) and MYA-4's (*Bison*) and the new French Mirage IV's, all of which had been designed for high altitude operations, prepared to go to war at low altitude. The long-range bombers which were to have succeeded them, all of them high-flying supersonic aircraft, were cancelled either in the design stage or after they had flown as prototypes.

At the time of the change-over from high altitude to low altitude penetration tactics, the mere fact that an aircraft was flying close to the ground was sufficient to confer upon it a high degree of immunity from radar-aimed missiles and guns, and radar controlled interceptions. As a result, electronic countermeasures became less important in assisting the bomber through the defences: if the defending radar operators could not pick out the attacking aircraft from the mass of ground 'clutter' on their screens, why bother to jam the sets? Indeed, if the jammers were switched on at the wrong time, the defenders might gain some warning of the proximity of bombers they would not otherwise have. The long-range bombers, the only aircraft to require a comprehensive range of electronic countermeasures devices, now had little need for them; after a second 'great leap forward', the science once again fell out of favour.

Re-Inventing the Wheel

It seems that every time we go to war, we have to re-invent the wheel.

– A US AIR FORCE OFFICER, speaking during a symposium on electronic warfare.

The hiatus in the development of new electronic counter-measures equipment was not to last long. The war in Vietnam, which had been simmering since 1945, began to escalate rapidly during the 1960's as US forces were committed in progressively greater numbers. In February 1965, US Air Force and Naval aircraft began regular attacks on targets in North Vietnam.

Initially the North Vietnamese air defences were weak, comprising a small number of anti-aircraft guns, some with radar control, and a few outdated fighters. The fighters and fighter-bombers involved in the airstrikes, for the most part F-100 Super Sabres, F-105 Thunderchiefs, B-57 Canberras, F-4 Phantoms and A-4 Skyhawks, approached their targets at altitudes around 10,000 feet where they were above the effective reach of the ground fire. When they descended to make their bombing runs, they relied on their speed and manoeuvrability to keep down losses. For a while, it would be sufficient.

Then, in April 1965, came an ominous development. A Naval RF-8 Crusader reconnaissance aircraft returned to its carrier, the *USS Coral Sea*, with photographs of new con-structional work some 15 miles to the south-east of Hanoi. To those who had followed the evolution of the Russian air defences, the scenes depicted on the prints were familiar enough: they showed a site being prepared to take an SA-2 guided missile battery. During the months that followed several similar sites were photographed under construction in North Vietnam; and on 24th July the missiles knocked down their first victim during the conflict, an Air Force Phantom.

Representative of the early missile engagements, and the first to involve US Navy aircraft, was that on the night of 11th August. A pair of Skyhawks from the attack unit VA-23, operating from *USS Midway*, was flying at 9,000 feet some 60 miles to the south of Hanoi on an armed reconnaissance mission. The leader, Lieutenant Commander D. Roberge, observed what appeared to be two bright patches of diffused light wavering beneath cloud about 15 miles to the north of him. The lights wandered closer and closer until suddenly they emerged from cloud. Only then did their deadly peril become clear: the 'lights' were in fact the flames from a pair of rockets hurtling towards the aircraft. The pilots threw their aircraft into steep diving turns, but already it was too late. A few seconds later the missiles exploded one after the other, peppering Roberge's aircraft with fragments and destroying his wing man's machine. Roberge managed to limp back to *Midway* in his scorched and wrinkled Skyhawk.

The missile system involved in these actions, designated by NATO the SA-2, was the same as that which had shot down Francis Powers' U-2 aircraft five years earlier. The SA-2 dated from 1954 or earlier and had been designed to engage aircraft flying at medium or high altitudes. Unveiled in November 1957 during a parade in Moscow, it became the first Russian surface-to-air missile system to be deployed in large numbers and, later, exported; it is still in service in large numbers.

The missile used with the SA-2 system is code-named *Guideline* by NATO and has appeared in several different versions. Typically it is about 35 feet long, weighs about 5,000 pounds at launch and carries a high explosive fragmenting warhead weighing 288 pounds; it has a maximum speed of Mach 3·5 and its maximum range is about 25 miles.

The missile is a two-stage weapon, with a solid fuel booster rocket to provide high initial acceleration after launch, and a liquid fuel sustainer motor. *Guideline* is a command-guided missile; that is to say it is directed to its target by radio-command signals from the ground.

The SA-2 control radar has the NATO designation *Fansong* and, like the missile itself, it exists in several different versions. In one of its early forms the *Fansong* operates on frequencies in the 3,000 megacycle (10cm wave-

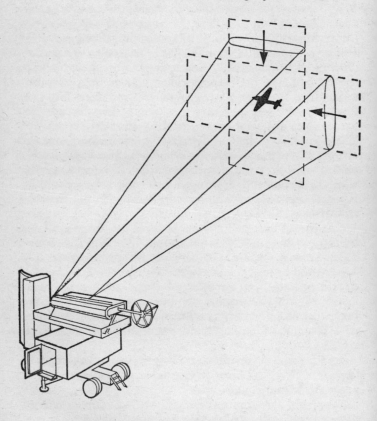

Diagram of 'Fansong' Radar Beams

length) band. The radar uses two trough-shaped scanners mounted on top of the operating cabin: the horizontal trough radiates a thin fan-shaped beam which sweeps rapidly from side-to-side through an arc of 10 degrees; the vertical trough radiates a similar fan-shaped beam which sweeps up-and-down, also through an arc of 10 degrees. At a typical engagement range, 10 miles from the radar, the beams sweep a square area of sky about two miles wide and 10,000 feet deep. Mounted to one side of the horizontal trough aerial is a dish which radiates the command signals to the missiles.

The usual type of *SA-2* site has the *Fansong* radar positioned in the centre and six *Guideline* launchers spaced evenly round it in a circle of radius 80 yards. Because of its narrow scan pattern, the *Fansong* has to be assisted to locate its targets by a search radar and a height-finder situated nearby. When a target has been allocated to them, the *Fansong* crew traverse their operating cabin and the aerials mounted on top of it until it is aligned on the target's bearing; then the entire aerial system is tilted until the target is in the centre of the scan pattern. The *Guideline* missile launchers selected for firing traverse with the operating cabin. Once the *Fansong* operators have the target in their field of view, they track it while their firing computer works out a point in the sky ahead of the target, at which the missiles should be aimed in order to score a hit. This done, provided the target is within firing range, the *Guidelines* are launched in salvoes of two or three.

After launch, beacons fitted in the missiles send back signals so that the computer can keep track of their movements. The impact point is continuously re-calculated and guidance instructions are radiated to each missile in turn to maintain them on a collision course even if the target evades. When the missiles reach the target they explode on impact, when the proximity fuse detects a target nearby, or on command from the ground.

It has been necessary to describe the SA-2 missile system in some detail, to enable the reader to understand the many possibilities there are for countering it. The simplest in theory, if not in practice, is to route aircraft so that they do not fly within range of the *SA-2* batteries. Or, if they have to pass near to the missile sites in order to reach their targets, the aircraft can fly at low altitude so that the *Fansong* operators cannot track them for long enough to complete an engagement. A further alternative is to mount air attacks on *SA-2* batteries likely to be troublesome.

Moving on to the possibilities for jamming, the search radar and the height-finder directing the *Fansong* are worthwhile targets; without their help, the *Fansong* operators might be unable to find their target in time to complete an engagement before it passes out of range. This jamming can be radiated from aircraft orbiting outside the range of the missiles. Next in line for jamming or spoofing is the *Fansong* itself; if either the azimuth or the elevation beams are jammed the control computer cannot work out where to send the missiles. Or large amounts of Chaff might be used to blanket the sky over the radar so that continuous tracking of the aircraft is not possible. Even if the jamming is not effective continually, it may still be able to prevent the missiles from reaching their target. To engage a target flying 10 miles from the launcher, a *Guideline* requires a flight time of about 25 seconds. Add to this the initial tracking time prior to launch, and the operators have to track their target for nearly three-quarters of a minute from the start of an engagement until its end. Continuously effective jamming for any 15-second period during that time would probably be sufficient to save the intended victim.

Even if, despite everything, the men in the *Fansong* operating cabin are able to track the aircraft through the jamming, that does not necessarily mean that they can destroy it. It is possible, though difficult, to jam the radio link conveying the command signals to the missiles. Or the beacon transmission from the missiles themselves can be jammed, so that the control computer can no longer keep track of them. Or it might be possible to feed false ranging signals into the missiles' proximity fuses so that they detonate before they get anywhere near the target.

Any one of these countermeasures, if effectively carried

out, could prevent a salvo of *SA-2* missiles from destroying its target.

As can be seen there were, theoretically, many different ways of outwitting the *SA-2* missile system. But initially few of these were available to the American fighter-bomber pilots operating over North Vietnam. If they were to attack the more important targets they had to enter the missiles' engagement zones; and if they flew to and from the target at low altitude they were liable to suffer heavy losses from the increasingly effective light flak defences. Air attacks were launched on the missile batteries and several were knocked out, but the North Vietnamese soon became adept at moving their batteries around and camouflaging them. Stand-off jamming aircraft, EB-66 Destroyers and EA-6 Prowlers, provided useful support for the fighter-bombers by jamming the search radars and height-finders working with the missile batteries.

Aircrews soon found that if they flew at 7,000 to 10,000 feet a correctly-timed evasive manoeuvre enabled them to outwit the missiles. As they saw the missiles being launched they would turn towards the launching site and dive steeply towards the ground; when the *Guidelines* completed their boost phase, curved downwards and tried to follow, the pilots would pull their aircraft up inside the arc of the missile. This presented the *Guidelines* with a difficult manoeuvring target close to or beyond their engagement capability. The manoeuvre worked well enough unless cloud concealed the missiles until it was too late. Fortunately for the crews, however, a technically simple addition to their aircraft could overcome this problem: a radar warning receiver. Under a crash programme a special receiver, designated the APR-25, was fitted into the fighter-bombers. This provided crews with warning of when their aircraft were being tracked by enemy missile- or gun-control radars and gave an indication of the direction of the threat. The pilot could hear the distinctive rattle of the *Fansong* signal in his earphones, with a change in pitch as the enemy operators switched their radar from the search to the engagement mode. An additional receiver, the APR-26, gave an indication of when the missile command signals were being radiated (indicating that a salvo of *Guidelines* was on its way).

As the defences strengthened and more and more missile sites became active, it became clear that alone the evasive manoeuvre did not always provide sufficient protection: fighter-bombers manoeuvring to avoid a salvo of missiles from one site might position themselves nicely for a salvo from another. The solution was for the fighter-bombers to carry jammers for self-protection, though these were less easy to produce rapidly than the receivers. Previous jamming devices had almost invariably been carried on board large aircraft where space was not a problem and there was usually a specialist electronic warfare crewman to operate them. The new family of jammers had to be small enough to be carried by a fighter-bomber and simple enough to be operated by a pilot in addition to his other tasks. Most of the fighter-bomber types were already brimful of equipment and had little room inside the airframe for major new additions; so it became usual to fit the jammers in a 'pod' mounted on a wing or fuselage bomb station.

Under a series of crash programmes during 1967 and 1968 several different types of jamming pod were built and shipped to south-east Asia for operational trials. One of the first such pods to go into large scale production was the Hughes ALQ-71, over 700 of which were built. This pod was 10 feet long, 10 inches in diameter and weighed 330 pounds; it carried six separate jamming transmitters, two covering each of the three frequency bands used by the SA-2 missile system, each feeding a vertical aerial mounted on the underside of the pod.

Jamming pods soon became very clever. The ideal from the operational point of view, though it did raise the level of complexity, was for pods to carry responsive jammers. Such a jammer was fitted with a control receiver which, when it picked up signals from a threatening enemy radar, tuned in the jamming transmitter to that frequency and, if programmed to do so, switched on the jamming automatically. From time to time the jamming was switched off to allow the receiver to hear whether the enemy radar was still trying to track the aircraft. A small memory unit fitted to the receiver enabled the radar signals to be compared with known signals from enemy sets, and identified. Thus the jammer could be given a list of priorities so that it could, for example, cease jamming an enemy search radar and concentrate all of its power on a missile control radar posing a more immediate threat. Such a receiver could also tailor the jamming transmissions, to counter in the best way the particular type of radar encountered.

A further electronic warfare innovation during the Vietnam war was the formation of special fighter-bomber units to engage enemy surface-to-air missile and gun batteries. Nicknamed *Wild Weasels,* these units provided cover for other fighter-bombers attacking their targets. Initially equipped with the two-seat version of the F-100 Super Sabre, the *Wild Weasels* soon re-equipped with the two-seat F-105G Thunderchief. The electronic warfare operator in the rear cockpit used a special receiver which enabled him to take precise bearings on one enemy radar out of perhaps several radiating in the area, and direct the pilot towards it. The main offensive armament of the *Wild Weasel* aircraft were the Shrike and the Standard anti-radiation missiles, which homed on the signals from the selected enemy radar.

The reader may get the feel of a typical *Wild Weasel* mission from this account by Captain Don Carson of the 44th Tactical Fighter Squadron, who flew the F-105G in this role from Korat in Thailand early in 1968.[1] During this night action Carson was to work with a second aircraft against enemy surface-to-air missile or gun batteries, while other F-105's attacked a munitions storage area near Dong Hoi.

As we crossed into North Vietnam, I commented on the excellent visibility and brightness of the full moon. I could see the reflections from the rivers and small lakes as we headed toward Packard's [callsign of one of the aircraft attacking the storage area] target run-in.

1. The account first appeared in *Air Force Magazine,* April 1974.

We trolled back and forth along his target area, listening and looking for an enemy SAM [surface-to-air missile] or AAA [anti-aircraft artillery] activity. This was not unlike trolling for fish, except this time we were the bait. A couple of strobes from a radar-guided gun and a low pulse rate frequency [PRF] SAM radar light indicated that someone knew we were there. They probably also knew that since we were alone and carried no jamming pods, we were a Weasel bird.

SAM operators normally did not bother a Weasel unless they could be assured of a good shot at him. If they fired and missed, they gave away their position and risked destruction. Even if they did not launch a SAM we could fire a Shrike to ride down their radar beam if they stayed on the air too long. They knew this and only turned their radars on for a few moments at a time . . . just long enough to keep track of our location. It was a game of cat and mouse . . . much like the way a real weasel hunts its prey.

We purposely turned our tail towards the SAM site that had been giving us the once over, hoping to get him to turn on his radar long enough for us to turn into him and launch a Shrike. No luck. He was too smart. He probably knew that the second Weasel bird, [radio callsign] Muskrat, was also in the area just south of our position. We had co-ordinated with Muskrat that we would stay north of the final run-in from the IP [initial point] and they would stay south.

The Raider birds were to hit storage and truck staging areas just north-west of Dong Hoi, in the southern panhandle of North Vietnam. We really did not expect too much SAM activity since it had been quiet for the past few nights. I checked the clock, and we had ten minutes until we were to be at the IP. Heading west, we descended to our briefed altitude and waited for Packard to check in.

'Vampire, this is Packard . . . on time', they called. I answered 'Roger, Packard, we will be at the IP in five'. I pushed the throttle up to get 550 knots as we turned to make the run-in. I briefed Packard on the SAM and AAA activity so they would have an idea of where to look for trouble.

Once again we flew the route to the target area, this time with Packard about two miles off our right wing and 2,000 feet low. The vector gear [direction-finding equipment in the Thunderchief] now began to show some increased activity. The North Vietnamese gunners and SAM operators would love to get off a shot at Packard, but they were wary of our Shrikes.

Several 37 mm guns began to open up on us, but they were well off to our left. The shells arc skyward and explode with a bright flash that is gone in a moment, leaving small puffs of smoke that are visible in the moonlight. They look almost beautiful, like roman candles, when they are far away from you. Don [the electronic warfare operator] comments on their bad aim tonight. I answer 'I hope it stays that way'.

'SAM . . . low PRF . . . two o'clock', Don calls. I can tell by his voice that it is not yet much of a threat. When the danger is more immediate, the pitch of his voice lets me know. When you are flying with someone every night, you learn to grasp the meaning of every inflection and sound.

About two minutes out from the target, we begin to really get some activity from the AAA gunners. They seem to know our exact location and altitude. They surrounded our aircraft with 37- and 57-mm rounds. Packard, off to our right, was getting it even worse. This was barrage AAA and not radar-guided, as we did not get any significant radar strobes.

Just to make things more interesting, we picked up a strong new SAM signal at ten o'clock. Don called out 'Three ringer at ten', indicating that he was getting a strong signal off to the left. The red warning lights in the cockpit and the rattlesnake growl in my earphones confirmed the higher pitch in Don's voice. Several times before I had thought we had a SAM launched at us when we didn't. Lightning and static electricity in weather could trigger the warning gear and give a momentary false indication of a launch. However, when it is for real, there is no doubt.

The SAM operator was serious, and he wanted Packard. The signal continued as I manoeuvred into position to launch a Shrike. We turned directly towards the site, telling Packard that we had a SAM locked on at his ten o'clock. The guns continued to hammer away at us. Pulling up into the SAM site, I launched a Shrike.

I had heard a hundred times that you should close your eyes when launching a missile at night to protect your night vision. This was like telling a boy at a country fair peep show that he would go blind if he looks at the dancing girls. I decided to risk one eye!

The Shrike lit off with a roar and left the F-105 with a burst of speed and a trail of brilliant fire. It was beautiful, but I did not have time to watch it for long. Now every gun in that part of the world had opened up on us.

I waited for Don to call the SAM launch, hoping the Shrike would get there first. The Shrike guided, and as we saw it impact, the SAM signal suddenly ceased.

'I think we got him', I called to John and Stan in Packard, who were approaching their bomb release point. John had to hold a precise heading and speed in order for Stan to get a good run-in for his radar bombing. The SAM launch would have prevented their making a run and might well have ended their night. I don't know why the SAM site waited to launch, but I am glad it did!

'Packard's off left.' We started a turn back to the west. They had carried a load of 750-pound bombs that were now on their way to the fuel and ammunition storage area.

The bombs exploded seconds later with a brilliant flash, followed by secondary explosions of bright orange. The fireballs shown brightly in the darkness.

We turned to escort Packard out of the area and pick up our next Raider. The guns were still hammering away but we were climbing above them as we headed back to the IP.

The second run-in with Buick was not quite as interesting as the first. Evidently we had knocked out the SAM radar, or at least scared them into shutting down for a while. We heard not one bleep from their radar the rest of that night. The AAA gunners still had a good supply of 37- and 57-mm rounds, but were not up to their usual level of accuracy.

We stayed in the target area for another hour escorting the remainder of Ryan's Raider aircraft in and out of the target area. They had gotten some good secondary explosions that night. There was a lot of fuel and ammunition that would never reach the Viet Cong in the south, thanks to their efforts.

As the North Vietnamese air defences became stronger MiG 17 and MiG 21 fighters, armed with infra-red homing missiles (NATO designation *Atoll*) and guns began to contest the American incursions more frequently. The defending fighters operated with the benefit of close control from radar stations on the ground, endeavouring to break up the attacks and force the fighter-bombers to jettison their loads clear of the target. In reply, the attacking fighter-bombers received increasingly strong escorting forces of F-4 Phantoms armed with Sparrow and Sidewinder air-to-air missiles and, later, guns. For the American fighter crews, operating as part of a numerically superior force over enemy territory, a major difficulty was that of identifying the contacts which appeared on their radar screens beyond visual range. The identification transponder fitted in each aircraft could establish it as a 'friendly', but the absence of identification signals certainly did not establish that it was a 'hostile': the aircraft's transponder might have been damaged by enemy action, or it might simply have failed (that can happen). So the fighter crews were not permitted to open fire on aircraft not positively identified as 'hostile' and the Sparrow long range radar homing missiles, with a theoretical maximum engagement range of over ten miles, saw relatively little use. For the manoeuvring type of combat which developed, the shorter range Sidewinder infra-red homing missile was the more suitable. Both sides began using infra-red homing missiles in fairly large numbers; to counter this weapon, many fighters and fighter-bombers were modified to carry infra-red decoy flares to seduce such missiles away.

For the greater part of the air war over North Vietnam there existed the paradoxical situation where the strategic targets in the north were attacked by shorter-ranging fighter-bombers, while the long-range B-52 heavy bombers busied themselves against tactical targets in the south. To many, this writer included, at the time it seemed that the long range bombers were simply too large, too slow and too cumbersome to survive over North Vietnam.

We were proved wrong in April 1972, when the previous policy was reversed and B-52's began operating over the north. On the 8th a force of twelve of these heavy bombers attacked the railway marshalling yard at the coastal town of Vinh and returned without loss. As the defences of a target in North Vietnam went, those guarding Vinh were not particularly strong and in any case only a short penetration of the defences was necessary to reach the target. Nevertheless the attack did demonstrate that the heavy bombers, attacking by night at altitudes between 30,000 and 40,000 feet, could survive if they were given sufficient jamming protection.

For the North Vietnamese, it was merely a taste of what was to come. During the early morning darkness of 16th April, the heavy bombers mounted a considerably more ambitious operation. Covered by EB-66 jamming aircraft and F-105G *Wild Weasels*, a force of seventeen B-52's with fighter escort launched an attack on petroleum storage areas at the heavily-defended port of Haiphong. Preceded by two lead aircraft, the main force of B-52's flew in successive cells of three aircraft to provide mutual jamming protection. The tactics proved highly successful: none of the bombers or their escorting fighters was shot down even though large numbers of missiles were launched at them. After dawn a force of 32 F-4 fighter-bombers, similarly escorted by fighters and electronic countermeasures aircraft, attacked petroleum storage areas at Hanoi; simultaneously, a small force of Naval attack aircraft raided other targets in the area. Four MiG 21's attempted to engage the raiders, but they fell foul of escorting F-4's which shot down three of them. During these operations 242 *Guidelines* were recorded as having been launched at the various raiding forces and there was a large volume of ground anti-aircraft fire, much of it radar controlled. In spite of this only two aircraft were lost, a *Wild Weasel* F-105G and a Naval aircraft. It was a clear demonstration of the effectiveness of jamming to cover a set-piece operation mounted in the face of powerful ground defences; it showed, too, that ground launched missiles could be prevented from annihilating a force of heavy bombers flying at high altitude.

The lesson was driven home even more forcibly at the end of the year, during the series of large-scale attacks by B-52's against targets in the Hanoi and Haiphong areas. Designated *Linebacker II,* the operation began on the night of 18th December and continued with nightly attacks, with only a twenty-four hour pause for Christmas, until the 29th. In the course of the operation the B-52's flew just over 700 sorties; this meant an average of 64 heavy bombers, each with a 24-ton bombload, operating each night. Altogether fifteen B-52's (about 2 per cent of the sorties) and eight other aircraft were lost, all of the former and most of the latter to *SA-2* missiles. The raiders suffered most heavily during the initial attacks, the worst time being on the third and fourth days when six B-52's were destroyed. After that the defences began to sag, as the defending *SA-2* batteries were either knocked out by *Wild Weasel* aircraft or simply ran out of missiles which were not replaced as the storage areas were destroyed. During the raids which followed the Christmas bombing pause the heavy bombers were able to range over North Vietnam almost with impunity, having shattered the fighter and missile defences during the earlier attacks.

US Government sources have stated that the defenders fired between 750 and 1,000 *Guideline* missiles at the B-52's and other aircraft taking part in the *Linebacker II* strikes. Against jamming targets it was normal to launch the *Guidelines* in salvoes of three, so it would seem that during the eleven days the defenders tried to engage about 300 targets, or one-in-three of the B-52's and their supporting aircraft.

It is true that the defences were far more active during the first week of the attacks than later; but even if the maximum figure of 1,000 missiles had been fired during that period (which they were not), this gives an average nightly engagement rate of only fifty targets. The simple arithmetic of the released figures hardly squares with the stories of 'huge barrages of unaimed missiles hurled at the B-52's which appeared in the popular press. It seems far more likely that the *SA-2* crews did their best to control most, if not all, of their salvoes on to their targets, but were successful with only about one salvo in twenty. Considering that the heavily-laden B-52's had been flying at the ideal engagement altitude for the *SA-2* and could not manoeuvre to avoid them, it has been estimated that had there been no electronic counter-measures protection between 75 and 100 heavy bombers would have been lost during *Linebacker II*. Such a loss would have been too heavy to countenance and such a prospect would certainly have prevented the large-scale use of B-52's over North Vietnam.

The full story of the B-52 attacks on Hanoi and Haiphong has yet to be revealed but one thing appears clear: the tactics employed by the heavy bombers bore a close resemblance to those used by the RAF night bombers during the early months of 1945. Consider the similarities between the *Linebacker II* attacks and that on the Bohlen oil refinery on 21 March 1945 already described: the stream of bombers keeping together to swamp the defences, flying at high altitude and radiating jamming for mutual protection; the patrols by escorting fighters in front of the bombers and on their flanks to ward off defending fighters; the use of specialized jamming support aircraft, orbiting outside the range of the enemy ground defences, to neutralize the defensive radars. The B-52's flew three times as fast and nearly twice as high as a Second World War night bomber, and each carried between three and eight times the bomb load; but essentially their tactics were the same.

During the course of the air war over North Vietnam there had been a steady drop in the effectiveness of the *SA-2* missile system, as the various countermeasures took effect.

When it was first used on a large scale, in 1965, the *SA-2* destroyed about ten fighter-bombers for an estimated 150 *Guidelines* launched: an average of one kill for every fifteen missiles. By November 1968 one aircraft was being shot down for every 48 missiles fired. During *Linebacker II* one aircraft was destroyed for roughly every 50 *Guidelines* fired. In the words of one USAF officer commenting on the various measures to counter *SA-2*: 'I guess we must have been doing something right!'

Linebacker II did prove that a powerful barrage of electronic jamming, combined with vast quantities of Chaff and carefully evolved anti-missile tactics backed by *Wild Weasel* attacks on the launching sites, could indeed reduce an air defence system well-equipped with surface-to-air missiles to near-helplessness. It should be remembered, however, that over North Vietnam the USAF and US Navy planners had to deal with only one type of surface-to-air missile system, the *SA-2*; this had first entered service around 1956 and in spite of some major improvements it was getting decidedly long-in-the-tooth by the late 1960s. Similarly, the fire control radars used by the North Vietnamese anti-aircraft gun units were all obsolescent designs.

Within a few months of the end of the air war over North Vietnam there came dramatic proof – if anyone needed it – that Russian radar and missile technology had come a long way since the design of the *SA-2* had been finalized.

In October 1973 Egyptian and Syrian forces launched a massive pre-emptive attack to regain territory captured by the Israelis six years earlier. When on the first day the Israeli Air Force fighter-bombers arrived over the battle area, a rude shock awaited them. The pilots were familiar with the dangers of the *SA-2* system, and the *SA-3* which was considerably more effective against low-flying aircraft. But both of these were semi-static systems, not to be expected near the forefront of an enemy armoured thrust. As the fighter-bombers dived into the attack, several were picked off by missiles from *SA-6* batteries situated close to the battle front. Both the fire-control radar (NATO designation *Straight Flush*) and the launchers for the missiles (NATO designation *Gainful*) were mounted on tracked vehicles which allowed the system a high degree of mobility. More important, the *SA-6* had not been used in action before, it was still an unknown quantity and electronic counter-measures were not ready to meet it. In any case the *SA-6* system employs semi-active radar homing, which is intrinsically more accurate and more difficult to jam than the command guidance system used by the *SA-2*; in a semi-active system, the fire control radar on the ground sends out a beam which is locked on to the target, and the missile homes on to radar signals reflected off the target.

Nor was the *SA-6* system the only unpleasant surprise for the Israeli fighter-bomber crews. Complementing the missiles was the ZSU 23-4 anti-aircraft gun system, comprising four rapid-firing 23mm cannon with radar control all mounted on a single tracked vehicle. Several aircraft which had descended to low altitude to avoid the missiles, fell to these guns. Also used was the *SA-7*, a small man-portable heat-seeking anti-aircraft missile; not very effective against high speed fighter-bombers, this weapon was lethal against low speed aircraft or helicopters.

Between them, the *SA-6* missiles and the ZSU 23-4 gun batteries probably brought down the majority of the eighty-odd Israeli fighter-bombers destroyed during the first week of the war. Many other aircraft returned with damage which prevented their being used again during the conflict.

After the first week of fighting the Israeli losses in fighter-bombers fell dramatically; during the next eleven days of the war thirty-five aircraft were lost. What part did electronic countermeasures play in reducing losses to this level? The Egyptians, Syrians and Israelis, who are in a position to know, will not say; the journalists, who will say, are generally not in a position to know. There can be little doubt that after the initial blood-letting the Israeli Air Force became a good deal more circumspect about attacking ground targets and gave up the attempt to destroy the heavily defended pontoon bridges over the Suez Canal. Moreover, as the battle fronts stabilized, there was less need to risk the dwindling number of fighter-bombers in do-or-die attempts to halt enemy thrusts. When anti-aircraft missile or gun batteries were located close to the front line, old-fashioned ground artillery fire would have been effective in dealing with them. Also, considering the furious pace at which the battle was fought, it is likely that some batteries simply expended all their missiles and supply problems saved several aircraft.

The Iraelis have remained tight-lipped about their improvised methods to counter *SA-6*, probably because there is little to say. The war lasted only eighteen days, an almost impossibly short time in which to bring in anything effective.

This brings the story of electronic countermeasures to the present day. As the reader has seen, twice in the past this science has risen to a position of the highest importance only to be overtaken by events. Currently, in the mid-1970s, electronic countermeasures are again being developed at the highest priority for all the major air forces. So the question will be asked: is this yet another rise before a fall? Almost certainly the answer is 'no'. Electronic countermeasures are now here to stay.

Prior to the Vietnam war the use of electronic countermeasures was confined almost entirely to the long-range bomber forces, to aid aircraft to penetrate enemy defences. So the importance of jamming rose or fell according to the needs of the long-range bombers, which in any case constituted only a small proportion of the numerical strength of their respective air forces. The introduction of newer and better surface-to-air missiles and radar controlled anti-aircraft guns has changed the picture completely. No longer are such weapons relatively large and immobile and confined to home defence or the rear area of the battlefield: they must now be expected everywhere, including the forward battle area. Since these weapons have shown that they can be devastatingly effective against unprotected aircraft, some form of electronic countermeasures protection is necessary for *every* aircraft likely to have to operate within range

of enemy anti-aircraft guns or missiles, down to and including army helicopters. In the past electronic countermeasures
equipment had invariably to be squeezed into or on an aircraft as an afterthought, usually many years after the design
had been finalized; today such equipment is being designed
into the aircraft from the start.

Nor are the electronic countermeasures now confined only
to the air war. The sinking of the Israeli destroyer *Eilat* in
1967 by Russian-built *Styx* surface-to-surface missiles has
highlighted the importance of jamming protection for warships. Although ships are considerably larger and slower
than aircraft, the electronic countermeasures tools for their
protection are essentially the same: receiving devices to
provide timely warning of missile attack; jamming transmitters to confuse the radars directing the missiles or the
missiles themselves; and Chaff deployed explosively from
special rockets to provide alternative targets on which
missiles can home.

The main use of electronic countermeasures in the for-
seeable future, however, is likely to be in the air war.
One can get an idea of what is now possible from the scale
of equipment intended for the American Rockwell B-1
bomber. From its initial conception this aircraft was designed to carry a veritable powerhouse of jammers, comprising some seventy units positioned all over the airframe and
together weighing some 3,000 pounds. The transmitters fed
computer-controlled directional aerials which beamed the
jamming at the victim radars; if focused in this way, the
effectiveness of jamming can be considerably enhanced. In
addition to its jamming and deception equipment, the B-1
was designed to carry considerable quantities of Chaff cartridges and infra-red decoy flares. Each B-1 was to carry up
to twenty-four SRAM supersonic air-to-surface nuclear missiles; the bomber crews could thus hit their targets from
outside the range of enemy surface-to-air missiles or, if the
enemy missile belts were too thick for this, the B-1's could
use SRAMS to clear themselves a path through the defences.

There is no doubt that for the rest of this century the B-1
could have fulfilled the task for which it was intended: to

deliver missiles to the most heavily-defended targets in Russia. But in the end the B-1 was defeated, not by those defences, but by the sheer cost of the project. By December 1976 the projected cost of getting 240 of these bombers into service, including development, spares and training, stood at $22·9 billion: $95 million for each operational aircraft. The richest nation on earth baulked at this price tag and the US administration rejected the B-1, even after prototypes had flown, in favour of large numbers of the vastly-cheaper long-range cruise missile.

The lesson of the B-1 is clear enough: if military aircraft are to survive the expenditure reviews that must be expected in time of peace, unit costs have to be held down. This in turn means that the best solutions to problems, which in jamming as in almost any other field are usually the most expensive, will often not be open to designers. For the future we can expect jammers to become more and more adaptable, using miniaturized computers to direct the available power to greatest effect. By such means it is possible for one jammer to put out signals in rapid succession to jam several different radars spread over a wide part of the frequency spectrum. Mass-produced, such jammers need not be excessively expensive; nor will they be too large to fit inside the structure of the average attack aircraft.

One increasingly-attractive option, open to the designers of certain types of military aircraft, is the electronic countermeasures technique known as 'stealth'. By careful shaping of an airframe during the initial design stage, it is possible to build an aircraft with an extremely low radar signature. One way, of course, would be to cover the aircraft with some form of material that would not reflect radar signals at all; but although they certainly exist, effective radar-absorbent materials tend to be rather thick and they are not strong enough to resist the forces of air passing over them at supersonic speeds. By carefully shaping the airframe, however, much the same aim can be achieved without this penalty. A radar signal bounces off an aircraft in the same way as a tennis ball bounces off the ground or a racquet; and unless the signal, or a sufficiently great proportion of it, is reflected back in the direction of the radar, the aircraft simply will not appear on the screen. If the bottom and sides of an aircraft are correctly shaped, a very large proportion of the signals coming from below at any angle apart from the vertical can be deflected away from the radar – in the same way as an ace service in tennis bounces off the ground away from the server. If the engine air intakes and exhausts, both of which are great reflectors of radar signals, can be positioned out of view of a ground radar, say above the wings or the fuselage, an additional reduction in the aircraft's radar signature can be achieved. Obviously, the opportunity to employ 'stealth' techniques will depend a great deal on the role of the aircraft: it will be much easier to apply them to a reconnaissance aircraft than, say, an attack aircraft expected to carry a load of bombs under its wings. Even where 'stealth' techniques do not prevent an aircraft from being seen on radar, however, they make protection easier using other forms of countermeasure: if the size of the radar signature is reduced, smaller jamming or deception power levels will conceal it on the enemy radar screens.

Yet before one succumbs to the temptation to get carried away with the sheer technical brilliance of some of the electronic countermeasures, one should remember that fighter pilots and ground gunners were shooting down aircraft long before they had radar to help them. If a fighter pilot is sufficiently well trained and determined, he will not despair just because some of his own electronic gadgets are rendered unusable. Instead he may bore on in true von Richthofen style, finding his prey visually and hacking it down from short range using old-fashioned guns and a simple optical sight. If the opposing weaponry is primitive, there simply is nothing that can be jammed. Some commentators, usually out of ignorance, have tried to sell the line that electronic countermeasures are a panacea counter to all forms of defence. This is not so. A magnificent display of jamming will certainly reduce losses, but it will never render attacking aircraft completely invulnerable.

Electronic countermeasures can be regarded as a form of camouflage in a different dimension. Camouflage will rarely prevent detection altogether, especially if one has to move around to achieve anything useful; but it will slow detection perhaps until it is too late for the enemy to get in an effective blow. For a defensive system trying to stop aircraft penetrating to targets at low altitude at speeds of the order of ten miles per minute, or at high altitude at twice that speed, time is of the essence. By imposing delays in the operation of the defences at the critical time, electronic countermeasures can greatly reduce their effectiveness. The resultant guarantee of survival is not a firm one, but it is the best today's bomber crewman is likely to get.

Code-names

ABC 'Airborne Cigar'; airborne transmitter which jammed the German fighter control frequencies.

AI Airborne Interception; night-fighter radar equipment.

ALQ Prefix designation for USAF or USN electronic jamming equipment. For example ALQ-71 stands for Airborne, Countermeasures, Special Purpose equipment 71, a jamming pod.

APR Prefix designation of USAF or USN receiver equipment. For example APR-26 stands for Airborne, Radar, Receiver equipment 26, a radar warning receiver for aircraft.

'Aspirin' Jammer to counter the German *Knickebein* navigational aid.

'Benjamin' Jammer to counter the German *Y-Gerät* bombing aid.

Bernhard German ground-to-air communication system.

'Boozer' Radar receiver fitted to RAF bombers.

'Bromide' Jammer to counter the German *X-Gerät* bombing aid.

'Carpet' US jammer to counter the German *Würzburg* fire control radar.

'Chaff' Originally the US code-name for metal foil to confuse radar, now in general use.

Chain Home British early warning radar.

'Domino' Jammer to counter the German *Y-Gerät* bombing aid.

Düppel German name for metal foil dropped to confuse radar.

'Fansong' NATO code-name for the controlling radar used with the Russian SA-2 guided missile system.

Flak *Fliegerabwehrkanon*; anti-aircraft gun.

Flensburg Radar receiver, to enable German night fighters to home on to the emissions from the 'Monica' tail warning radar.

Freya German early warning radar.

'Gainful' NATO code-name for the type of missile used with the Russian SA-6 guided missile system.

GEE British navigational aid.

Giant *Würzburg* German fighter control radar.

GLIMMER Feint attack on Boulogne, in support of the invasion of France.

'Guideline' NATO code-name for the type of missile used with the Russian SA-2 guided missile system.

HEADACHE Generic term for the measures taken to jam the German *Knickebein* navigational aid.

Heinrich Jammer to counter the British *GEE* navigational aid.

Himmelbett German system of controlling night fighting.

H2S British bombing aid.

H2X US bombing aid, similar to the British *H2S*.

Jagdschloss German fighter control radar.

Kleine Heidelberg German aircraft detection system, using radiations from British ground radar sets.

Knickebein German navigational aid.

Korfu German radar receiver which gave bearings on aircraft transmitting with *H2S*.

Lichtenstein German night-fighter radar.

Linebacker II US bomber attacks on Hanoi and Haiphong in December 1972.

Mammut German early warning radar.

'Mandrel' British jammer to counter the German early warning radar.

Mattscheibe German tactics to silhouette bombers flying above cloud.

'Meacon' Device to mask the radiations from German radio beacons.

'Monica' British tail warning radar.

'Moonshine' Device to produce a false picture on German radar equipment.

Naxos Radar receiver, to enable German night-fighters to home on emissions from the *H2S* radar.

Nürnberg Modification to the *Würzburg* gunlaying radar, to give some relief from 'Window' jamming.

'Oboe' British bombing aid.

'Perfectos' Device to enable British fighters to home on to emissions from the identification equipment of their German counterparts.

Postklystron Jammer to counter the British H2S.

'Quail' Radar decoy missile carried by and released from a B-52 bomber, fitted with reflectors to give it the same appearance on enemy radar as the parent aircraft.

SA-2 NATO designation for Russian surface-to-air missile system designed to engage aircraft flying at medium and high altitudes; non-mobile.

SA-3 NATO designation for Russian surface-to-air missile system designed to engage aircraft flying at low or medium altitudes; non-mobile.

SA-6 NATO designation for Russian battlefield surface-to-air missile system designed to engage aircraft flying at low or medium altitudes. Control radar and missile launchers fitted to tracked vehicles giving a semi-mobile capability (i.e. the system can be redeployed rapidly, but the missiles cannot be fired while the vehicles are on the move).

SA-7 NATO designation for Russian battlefield surface-to-air man-portable missile system designed to engage slow flying aircraft or helicopters at low altitudes. Shoulder or vehicle launched, fully mobile.

Seetakt German naval gunlaying radar.

'Serrate' Radar receiver, to enable British night fighters to home on emissions from the radar of their enemy counterparts.

'Shrike' ARM US airborne anti-radiation missile, used by 'Wild Weasel' units.

SN-2 German night-fighter radar.

'Standard' ARM US airborne anti-radiation missile, used by 'Wild Weasel' units.

'Straight Flush' NATO designation for the controlling radar used with the Russian SA-6 guided missile system.

'Styx' NATO designation for the Russian surface-to-surface anti-ship missile system fitted to fast patrol boats.

TAXABLE Feint attack on Cap d'Antifer, in support of the invasion of France.

'Tinsel' Scheme for broadcasting engine noises on the German fighter control frequencies.

Wassermann German early warning radar.

'Wild Weasel' Nickname of USAF aircraft equipped to engage enemy ground anti-aircraft gun and missile batteries.

Wilde Sau 'Wild Boar'; German tactics, to engage bombers over the target.

'Window' British name for metal foil dropped to confuse radar.

Würzburg German radar used to direct AA guns, searchlights and, for a short time, night-fighters.

Würzlaus   Modification to the *Würzburg* gunlaying radar, to give some relief from 'Window' jamming.

Y-Control   A method of controlling night-fighters using modified *Y-Gerät* equipment.

Y-Gerät   German bombing aid.

Z-Gerät   German bombing aid.

Zahme Sau   'Tame Boar'; German tactics, designed to bring night-fighters in contact with bombers moving to and from the target.

ZSU-23-4   NATO designation for Russian anti-aircraft gun system comprising four fully automatic 23-millimetre guns with radar control mounted on a tracked vehicle; fully mobile.

Index